MW01140623

DISCOVERING
PLANET EARTH

DISCOVERING PLANET EARTH

A GUIDE TO THE WORLD'S TERRAIN AND THE FORCES THAT MADE IT

GEORDIE TORR

This edition published in 2021 by Arcturus Publishing Limited
26/27 Bickels Yard, 151–153 Bermondsey Street,
London SE1 3HA

ISBN: 978-1-83940-868-7
AD008395US

Printed in Singapore

Contents

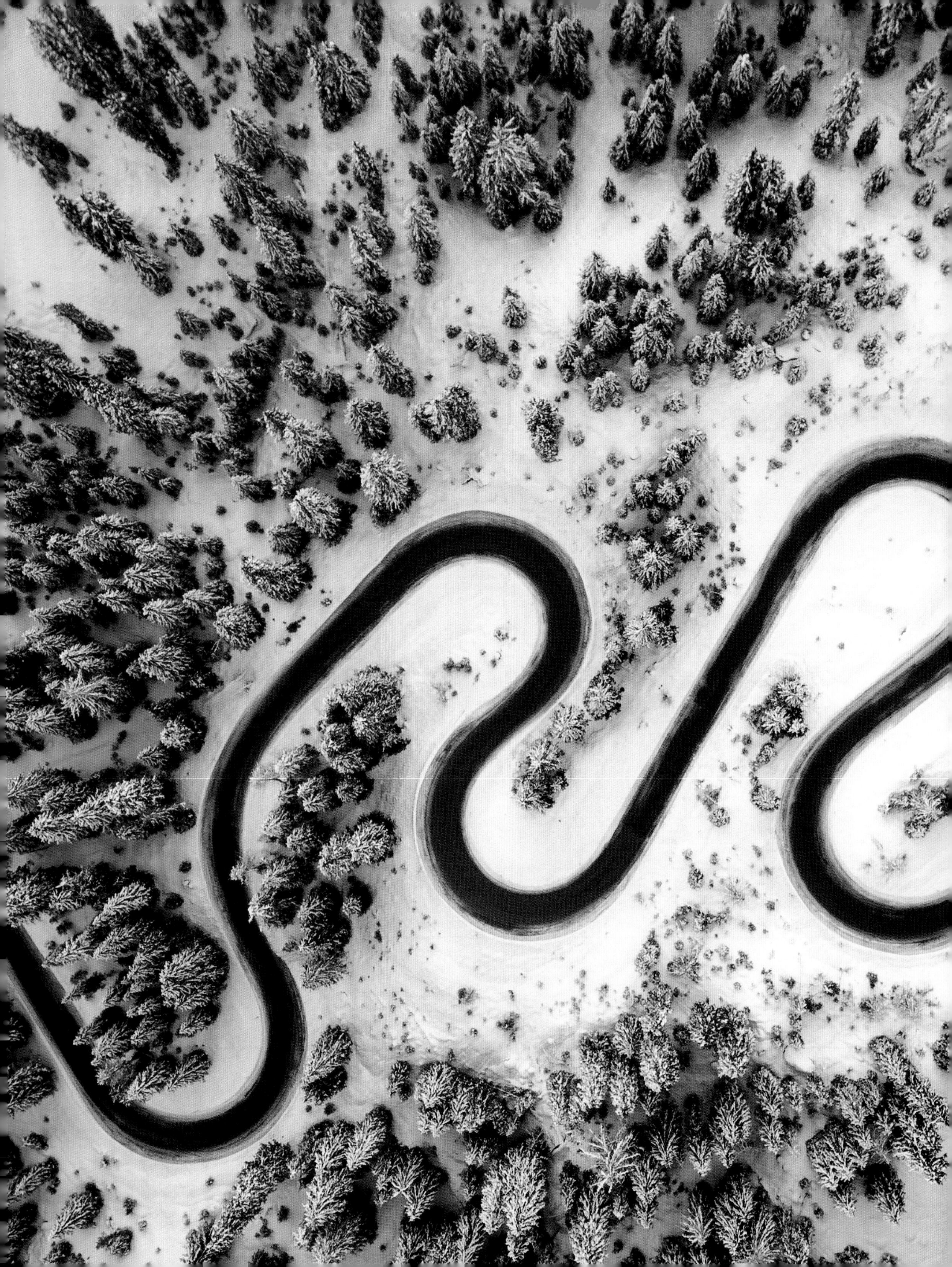

Introduction

From its searingly hot, dense core to its rocky surface and out to the limits of its atmosphere, the Earth is a restless, dynamic planet, forever shifting and changing.

The collision of continents creates vast mountain ranges and spawns volcanoes and devastating earthquakes; rivers and glaciers carve deep valleys and create broad, sweeping plains and fertile deltas; ocean currents and atmospheric circulation move enormous amounts of heat from the equator to the poles, influencing the climate and generating powerful storms.

This relentless reshaping of the Earth's surface is responsible for the remarkable diversity of its landforms and habitats. Falling rain carves out majestic caves and triggers landslides; ocean waves batter coastlines, sculpting them into elaborate forms; and strong winds blow desert dust halfway around the globe.

The Earth can be divided into three spheres—the land, the sea, and the air—each with its own characteristics, processes, and dynamics. While in some ways these different spheres are discrete, they are also, in many ways, interconnected, influencing each other in complex ways.

We are currently living through a period of rapid environmental change. Human activity has had a profound effect on the world's climate systems, altering them in ways that we're still struggling to understand. The consequences of this unprecedented act of planetary engineering are now playing out before our eyes in the form of retreating glaciers, melting permafrost, alternating droughts and floods, powerful storms, devastating wildfires, rising sea levels, and much more.

All of which means that there is no better time to discover the mystery and majesty of the planet we call Earth.

Below *Planet Earth seen from space.*

THE LAND

Although there are rock outcrops that date back to the Earth's early days, most of the land is dynamic and ever-changing, constantly being worn down by erosion, pushed up into towering mountain ranges by tectonic and other forces, and even newly created by volcanoes. Those same tectonic forces are also moving the land around the globe, the continents taking part in a slow, majestic dance—coming together to form vast supercontinents and then drifting apart again. Consequently, the land displays a remarkable diversity of forms: rivers and streams, and glaciers and icecaps have carved out deep valleys and canyons, and created fertile deltas; the ocean's power has battered coastlines, sculpting them into distinctive landforms. The land affects both our climate and our weather, playing a role in determining where rain falls and winds blow, where plants grow and deserts form. The Earth's total land area is roughly 58 million square miles (150 million square kilometers), or about 29 percent of its total surface. For much of human history, most of that land was wilderness, dominated by vast forests and grasslands. But over the past few centuries, we have changed the land beyond recognition, cutting down forests, planting crops, digging mines, building cities and much more.

The Carpathian Mountains, Ukraine. Covering an area of about 77,220 square miles (200,000 square kilometers), the Carpathians form the eastward continuation of the Alps. A geologically young mountain chain, they were relatively unaffected by glaciation during the last ice age and have mostly been shaped by running water.

// The origin of the Earth

The Earth formed out of a cloud of cosmic dust known as a solar nebula about 4.5 billion years ago through a process called accretion.

At some point, more than 4.6 billion years ago, static electricity caused particles of dust to begin to stick to one another, forming tiny objects known as particulates. As the particulates' mass grew, their gravity caused them to clump together with other particulates to form pebble-sized rocks that clumped together to form larger rocks, and so on. Eventually, this process of accretion led to the formation of tiny planets, about 0.6–6 miles (1–10 kilometers) in diameter, known as planetesimals.

The planetesimals collided to form larger bodies, one of which grew larger than the others and became the Earth. Over a period of some 120–150 million years, the nascent Earth was bombarded by more planetesimals, slowly enlarging further and further.

As the Earth grew, its gravitational attraction became stronger, drawing in more material and causing the

The formation of the Earth began with dust and small rock fragments sticking together until the resulting bodies, known as planetesimals, were large enough for gravity to become the dominant force. The protoplanet then grew swiftly, eventually becoming large enough for its surface to flatten out and an atmosphere to form. The final globe illustrated here shows the ancient supercontinent of Rodinia, which formed during the Precambrian period about 650 million years ago.

material that was already there to compress more tightly. Compression causes materials to heat up. Several other processes, including radioactive decay of elements such as uranium and collisions with comets and asteroids, made the Earth heat further, to the point where most of its constituent material melted and the planet was essentially a ball of lava floating in space. This caused a 'sorting' of the constituent parts, with the less dense silicate materials rising and eventually cooling to form the rocky exterior or crust, while the heavier, denser metals—mostly iron and nickel—sank to form the Earth's solid core. Materials with densities in between remained more or less molten, forming the intermediate layer, known as the mantle.

Gravity pulled the Earth into a roughly spherical shape. However, its rotation caused it to bulge slightly at the equator, forming what's termed an oblate spheroid (the Earth's circumference is 13 miles [21 kilometers]—or about 0.3 percent—longer around the equator than it is from pole to pole).

Above *Not long after the Earth formed, it was struck by Theia, a protoplanet about the size of Mars. Some of the debris from the impact went into orbit and coalesced to form the Moon, while much of the remainder rained down on the Earth.*

By about 4.5 billion years ago, the Earth had grown large enough that its gravitational field was strong enough to hold gas atoms to it, and it began to build an atmosphere (see page 130). Around this time, it was struck by a Mars-sized planet, known as Theia, whose metal core merged with the Earth's. The collision released an enormous amount of debris, which eventually coalesced to form the Moon, as well as a great deal of heat.

Further collisions with comets and asteroids over several million years deposited water on the young Earth's surface, while also creating deposits of metals and other heavy elements in the crust.

// The structure of the Earth

The Earth is made up of three main layers—the core, mantle, and crust—with very different compositions and behaviors.

The Earth's outermost layer, which accounts for less than 1 percent of its mass, is a rocky shell called the crust. It is rigid, brittle and cold compared to what lies beneath.

The crust is mostly made of the relatively light elements silicon, aluminum, and oxygen. There are two types of crust: oceanic and continental. Oceanic crust is younger than continental crust; it consists primarily of basalt that is continuously being created at mid-ocean ridges and destroyed in ocean trenches (see page 104). Continental crust, in contrast, is made up of a wide range of older igneous, metamorphic and sedimentary rocks, the most common of which is granite. Oceanic crust is denser than continental crust, causing it to sink lower into the mantle and thereby form the basins that house the Earth's oceans. When the two types of crustal material collide, it is the denser oceanic crust that is forced downwards.

Crustal thickness is highly variable: beneath oceans it may be as little as 3 miles (5 kilometers); beneath continents, as much as 50 miles (80 kilometers)—the thickest part lies under the Himalaya. On average, oceanic crust is about 4 miles (6.5 kilometers) thick and continental crust about 22 miles (35 kilometers) thick.

The next layer down is called the mantle. At close to 1,900 miles (3,000 kilometers) thick, it is the largest layer, comprising 83 percent of the Earth's volume. It is also relatively dense, making up about 68 percent of the Earth's mass. It consists mostly of oxides of iron, magnesium and silicon. In the upper mantle, the dominant rock is a mineral called peridotite.

The upper mantle consists of two layers: topmost is the cooler, rigid lithosphere, a region that includes the crust; below is the hot asthenosphere, which is semi-molten and hence capable of flowing slowly. On average, oceanic lithosphere is about 60 miles (100 kilometers) thick. It thickens as it ages and cools, adding material from below. Continental lithosphere is roughly twice as thick, although its thickness also varies. The lithosphere is broken up into a jigsaw puzzle of tectonic plates (see page 14).

Below the upper mantle lies the transition zone, where rocks neither melt nor disintegrate, instead becoming extremely dense. It's believed that this zone prevents

material moving into the lower mantle, a region of solid rock that is hotter and denser than the upper mantle. The intense heat of the lower mantle creates convection currents in the asthenosphere that help to move the tectonic plates around. In general, the deeper one goes into the Earth, the less detail is known for sure, and much from the lower mantle onwards is open to conjecture.

At the Earth's center is the core, which makes up about 30 percent of the planet and is almost twice as dense as

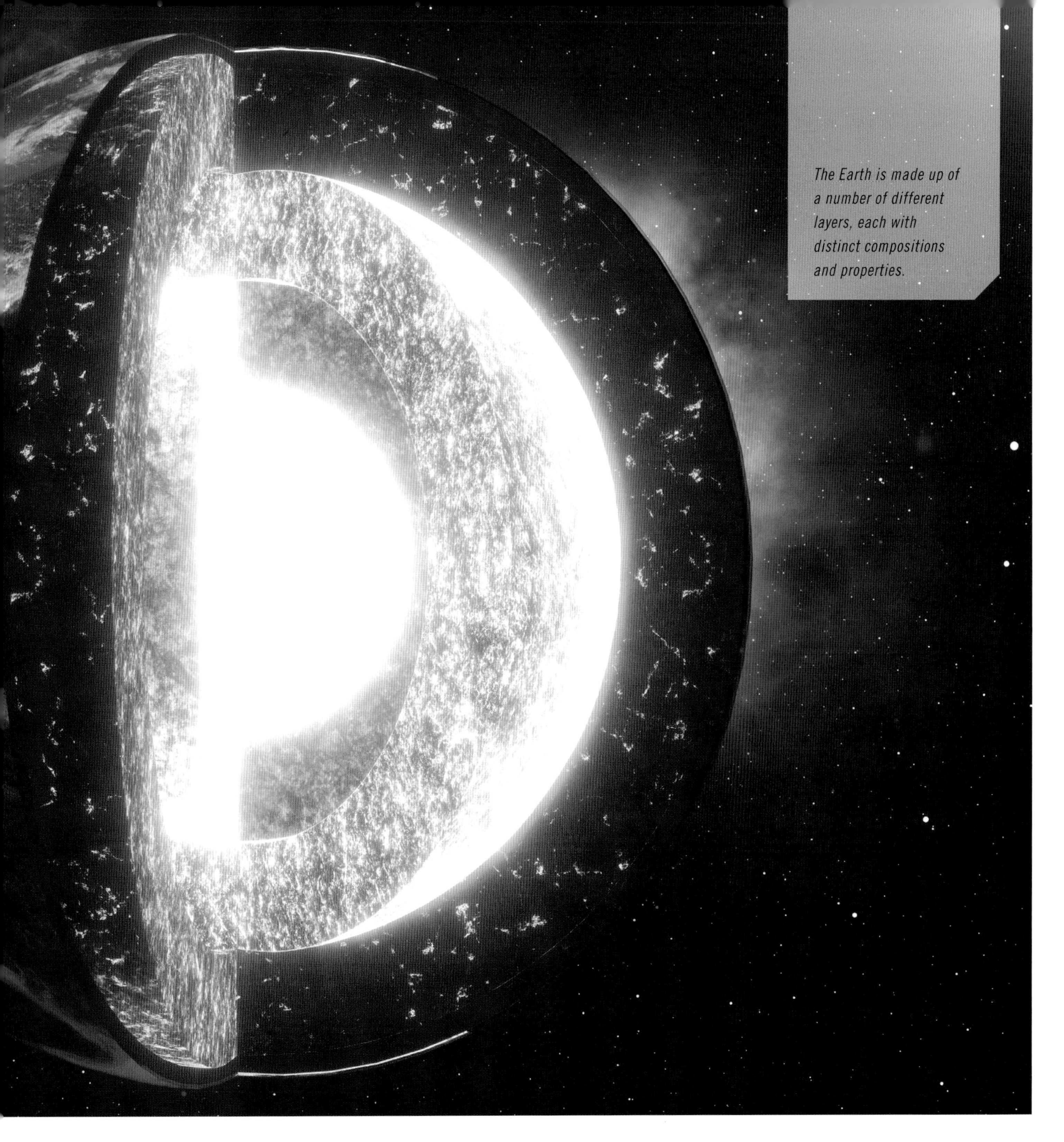

The Earth is made up of a number of different layers, each with distinct compositions and properties.

the mantle. The core is roughly 80 percent iron and 20 percent nickel, although a few other elements are also present, including gold, platinum, cobalt, and sulfur. It consists of two layers: the dense, solid inner core, which has a radius of roughly 760 miles (1,220 kilometers); and the liquid outer core, which is about 1,400 miles (2,200 kilometers) thick. There may also be an inner inner core that consists almost entirely of iron.

The temperature at the boundary between the inner and outer core has been estimated to be about 6,000°C (10,800°F), while the pressure is some 3.3 million times the atmospheric pressure at sea level.

Radioactive decay in the inner core, mostly of uranium and thorium, heats the outer core and keeps it liquid. It also churns the molten metal in huge, turbulent currents that generate electrical currents and, in turn, the Earth's magnetic field (see page 178). It's thought that the inner core spins slightly more rapidly than the rest of the planet.

// Plate tectonics

The Earth's crust is extremely dynamic, constantly being created and destroyed as a patchwork of rigid sections slowly shifts position. Over billions of years, plate-tectonic processes have changed the face of the planet, determining the positions of the landmasses and thereby influencing sea level and climate.

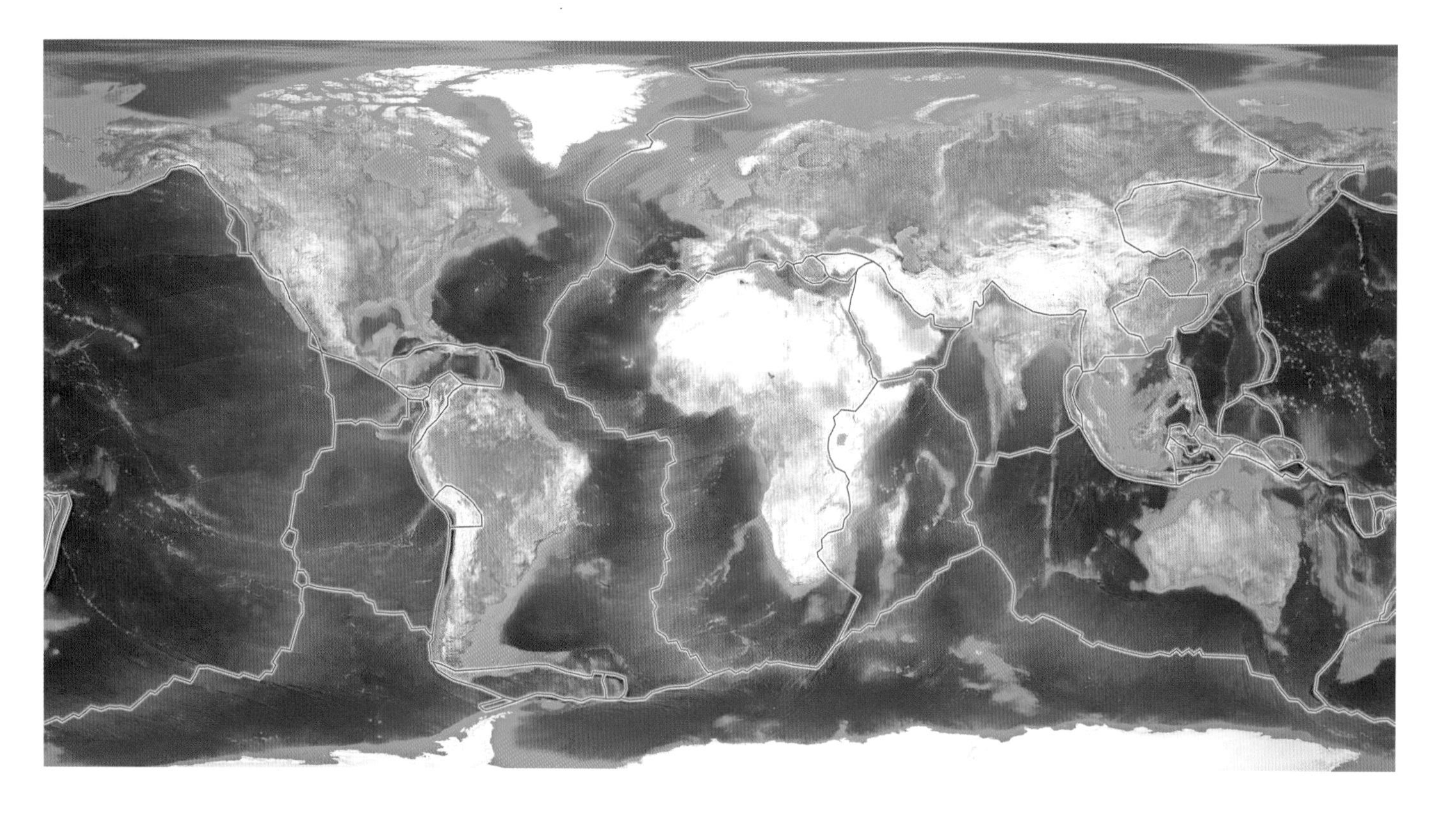

Above *The Earth's crust is divided up into seven major plates with areas of more than 7.7 million square miles (20 million square kilometers), about 15 minor plates with areas of 386,000–7.7 million square miles (1–20 million square kilometers) and dozens of smaller microplates.*

The planet's rigid outer shell, the lithosphere, is broken up into seven or eight major plates, named after the landforms that lie atop them: North American, Pacific, Eurasian, African, Indo-Australian (sometimes divided into Australian and Indian), South American, and Antarctic. There are also dozens of smaller plates. The crust that lies on a tectonic plate may be continental or oceanic; most plates contain both.

Tectonic plates are less dense than the material that lies below them in the semi-molten asthenosphere, so they effectively 'float' on and slide across it. They are all in motion relative to one another, at speeds of up to 4 inches (10 centimeters) per year, a process known as continental drift. The plates' boundaries may be convergent (moving towards each other), divergent (spreading apart), or transform (moving sideways in relation to each other). New crust is constantly being created at divergent plate boundaries (where material is extruded at mid-ocean ridges) and destroyed at convergent plate boundaries (where it is pushed down into the mantle in what are known as subduction zones).

The mechanism that underlies the plates' motion is still poorly understood, but it's generally agreed that two processes are involved: convection currents within the Earth's mantle and the "push" and "pull" caused by the creation and destruction of plates at their boundaries. The relative importance of these factors and their relationship to one another is unclear and much debated.

For some time, it was believed that convection within the mantle was the main driver of plate motion. Heat produced by the core creates convection currents in which material in the mantle, being hot, rises, spreads out horizontally as it approaches the crust and then sinks as it cools. Friction between this material and the undersurface of the lithosphere was thought to drag the plates around. However, scientists have been unable to identify mantle convection cells that are sufficiently large to drive plate movement.

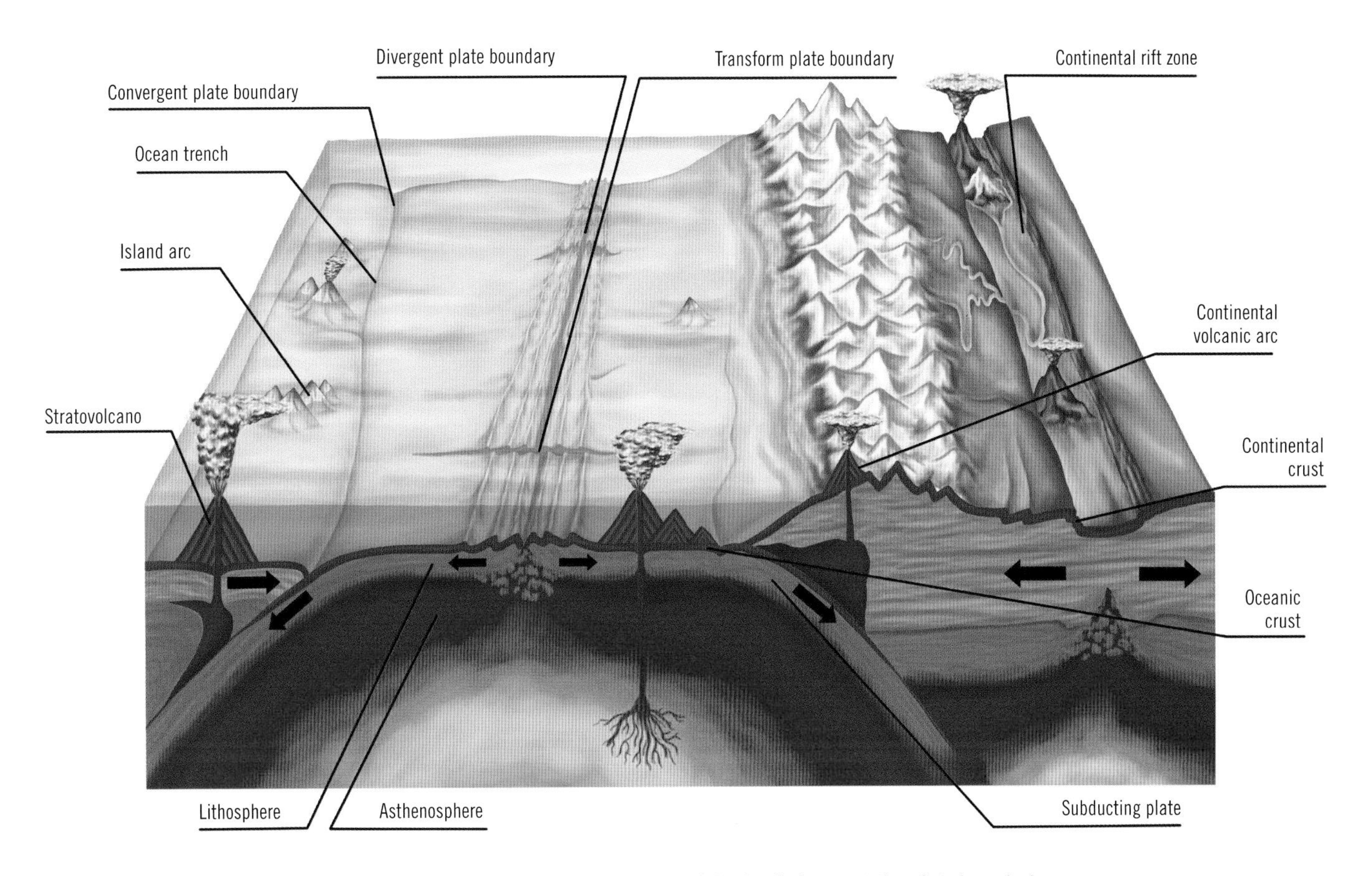

Above *The constant movement of the Earth's tectonic plates unleashes powerful seismic forces at the plate boundaries. At divergent boundaries, where the platea are moving awayf from each other, new crust is formed, while at convergent boundaries, one plate is forced down into the mantle, while the other may buckle to form massive mountain ranges.*

It's now generally agreed that plate movement is the result of what's known as "slab pull" and, to a lesser extent, "ridge push." As hot, newly formed lithosphere moves away from mid-ocean ridges, it cools, thickens and becomes denser, causing it to sink lower into the asthenosphere. Eventually, it reaches a subduction zone, where gravity forces it back down into the mantle. This process effectively pulls the plate away from the ridge and into the trench—hence the name slab pull. The problem with this theory is that, despite being in motion, the North American plate isn't being subducted. The same is true for the African, Eurasian, and Antarctic plates. Hence it's believed that gravity acting on newly formed plate material causes it to slide down, away from the mid-ocean ridges, pushing the plate in front of it and resulting in a ridge-push mechanism.

The drift of continental plates around the world has led to constant rearrangement of the landmasses. On a number of occasions, this has included creation of what are known as supercontinents (see over), when at least 75 percent of the crustal area at that time has come together to form a single landmass.

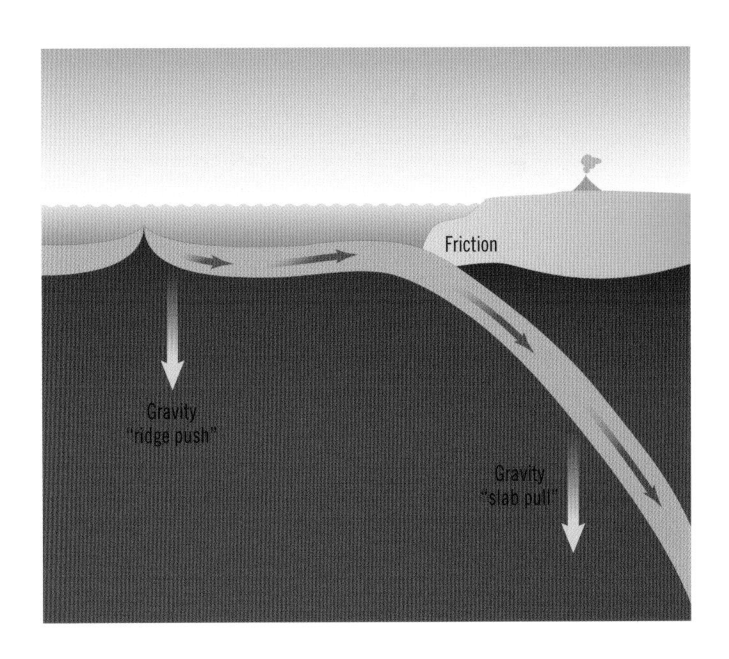

Above *The movement of the Earth's continental plates is thought to be driven largely by a mixture of "slab pull" and "ridge push," both of which are ultimately the result of gravity acting on the crustal material.*

Supercontinents

Since the plates began to move, between 3.5 and 3 billion years ago, supercontinents are thought to have formed and broken up every 500–600 million years or so. The most recent supercontinent, known as Pangaea, formed about 300 million years ago and appears to have included up to 90 percent of all the continental crust. It was surrounded by a global ocean named Panthalassa.

Pangaea began to break up about 215 million years ago. Starting about 200 million years ago, it split into two very large continents: Laurasia in the north (comprising what are now North America, Europe, and Asia) and Gondwana (today's southern continents, as well as the Indian subcontinent) in the south. These two continents were separated by a sea known as Tethys, the last remnant of which now forms the Mediterranean Sea.

Before Pangaea came Rodinia, thought to have assembled between 1.3 and 0.9 billion years ago. Rodinia appears to have lasted about 400 million years, before fragmenting about 760 million years ago. The giant world ocean that surrounded Rodinia is known as Mirovia. (Recently, scientists have proposed the existence of a short-lived supercontinent, Pannotia, said to have formed about 600 million years ago and broken up about 550 million years ago, but the idea remains controversial.)

As we go further back in time, evidence for supercontinents becomes more difficult to interpret; however, there is general agreement that about 2 billion years ago, a supercontinent, variously known as Nuna, NENA, Hudsonland, Hudsonia, Capricornia, Columbia, Midgardia, and Protopangaea, assembled. It's thought to have fragmented about 1.5–1.2 billion years ago.

Before that came Kenorland (also sometimes called Paleopangaea), believed to have existed around 2.5 billion years ago, which was itself preceded by Ur, thought to have existed around 3 billion years ago. The oldest proposed supercontinent, thought to have existed about 3.5 billion years ago, is known as Vaalbara.

Below *An artist's impression of the supercontinent of Pangaea.*

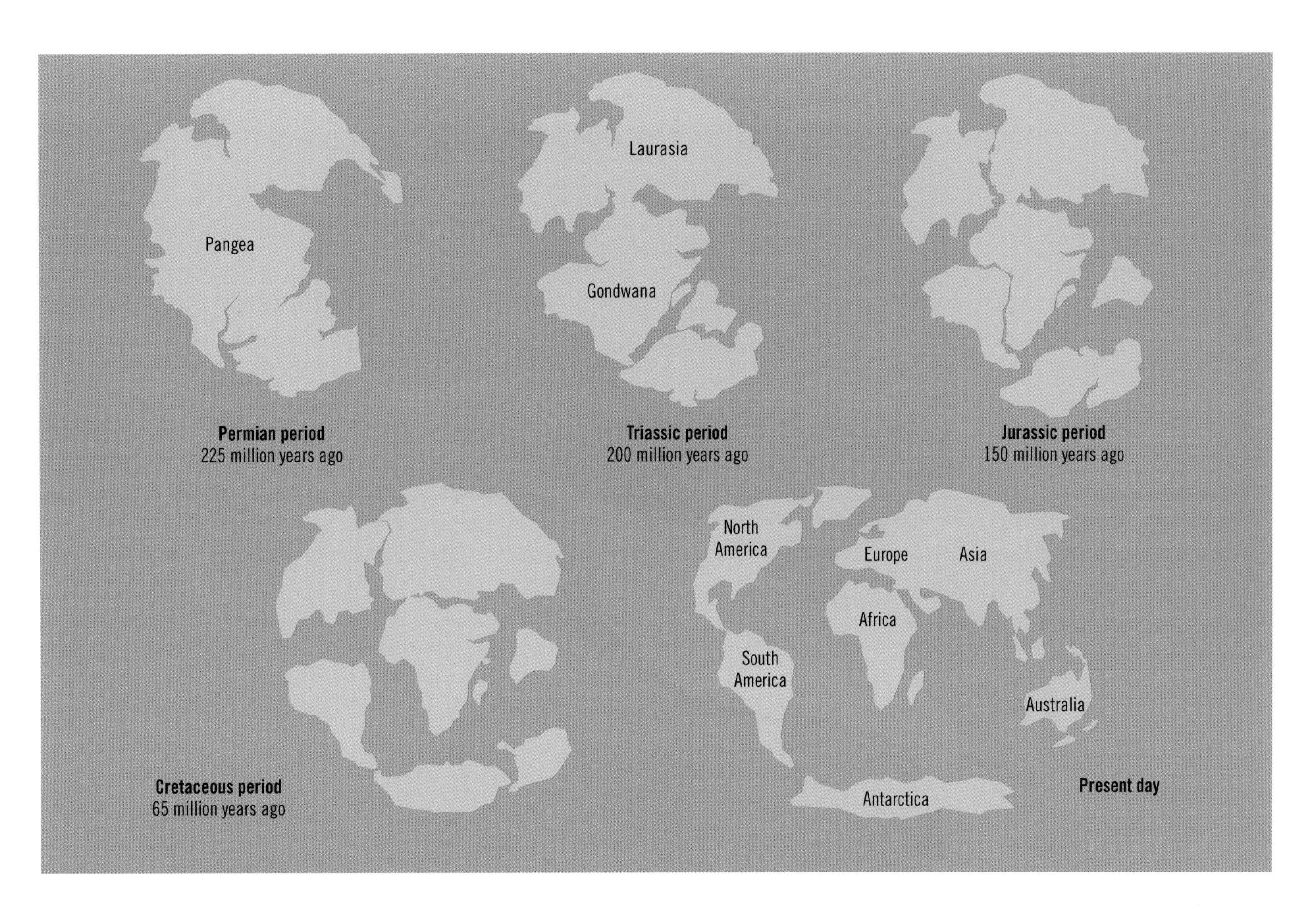

Above *The supercontinent of Pangaea began to break up during the Permian period, about 215 million years ago. It first split into two smaller supercontinents, Laurasia in the north and Gondwanaland in the south, before eventually separating into the landmasses that form today's continents.*

Running the cycle forward, scientists speculate that in about 50–200 million years, the Pacific Ocean will close up, with North America and Asia combining to form a new supercontinent that has been dubbed Amasia. Under this scenario, the Atlantic Ocean would expand into a new global sea.

The coming together and breaking up of supercontinents has had a significant impact on sea levels and ocean circulation patterns, which, in turn, have shaped global climate. For example, when Pangaea split into Gondwana and Laurasia, the formation of Tethys meant that the equatorial current could become circum-global. As the equatorial surface waters circumnavigated the world, they heated up, and some of this warm water appears to have made its way to the poles: Arctic and Antarctic surface-water temperatures were at or above 10°C (50°F), so the polar regions were warm enough to support forests.

Similarly, the break-up of Gondwana, with Antarctica moving south and becoming centered on the South Pole, and Australia and South America moving north, a new circum-global seaway developed around Antarctica, effectively isolating what became the Southern Ocean from the warmer waters to the north. Around the same time, the equatorial current system was blocked, which meant that the equatorial waters were heated less, the high-latitude waters cooled and an ice cap began to form on Antarctica.

CRATONS

At the centers of most of the Earth's continents are extremely old, thick, stable chunks of continental lithosphere known as cratons. Because oceanic crust is constantly recycled at subduction zones, pieces of sea floor never last more than about 200 million years. Continental crust, however, can be much older; in Greenland there are large chunks that are at least 3.8 billion years old. In some cases, the ancient crystalline basement rock is exposed, while in others, it's overlaid by sediments and sedimentary rock. Most of the world's diamonds come from cratonic areas. Pieces of continental crust that don't contain a craton, such as the island of Madagascar, are called continental fragments.

// Volcanoes

Ruptures in the Earth's crust through which hot rock, ash, and gas erupts, volcanoes have caused global famines and mass extinctions, and severely disrupted global climate.

Volcanoes are most common at convergent and divergent plate boundaries; most are underwater and most (roughly three quarters) lie on the Pacific Ocean "Ring of Fire." Those situated away from plate boundaries are typically above a so-called hotspot, thought to be caused by a plume of hot material rising through the mantle. As the plate moves over the hotspot, new volcanoes are created and older ones become inactive.

Volcanoes take two main forms. Shield volcanoes have a broad, gently sloping, shield-like profile, typically created when low-viscosity lava spreads a long way from the source before solidifying. This is more common in oceanic settings.

Stratovolcanoes exhibit the classic tall, steep-sided, conical shape. Examples include Mount Fuji in Japan and Mount Vesuvius in Italy. They are composed of layers of lava and tephra (see below); the lava is higher in silica, and hence more viscous, than lava from shield volcanoes, so it doesn't flow far from the vent. Because high-silica lavas tend to contain more dissolved gas, stratovolcanoes are more likely to exhibit explosive eruptions with great quantities of ash and

Below *Volcanic eruptions take several forms, each of which produce characteristic structures. Fissure vents are linear fractures through which lava emerges; shield volcanoes are typically created when low-viscosity lava spreads a long way from the source before solidifying, creating a broad, gently sloping, shield-like profile; stratovolcanoes are created when the erupting lava has a high viscosity and builds up around the vent as a tall, conical volcano; and lava domes are formed by a mixture of effusive eruptions of viscous magma and expansion due to magma being forced up below ground.*

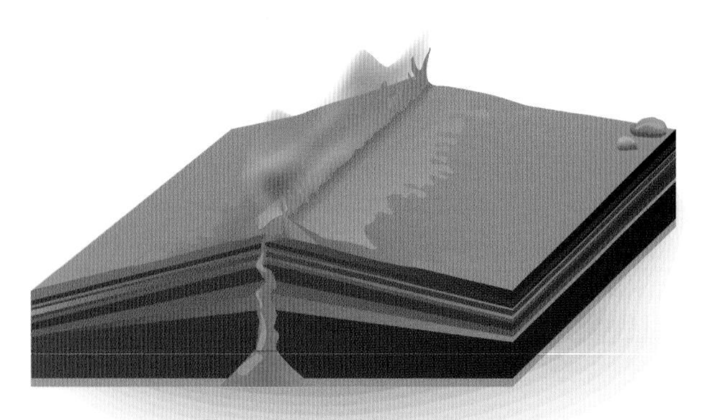

Fissure vent

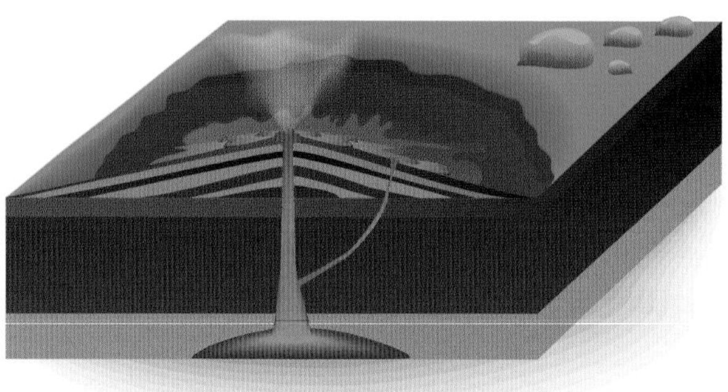

Shield volcano

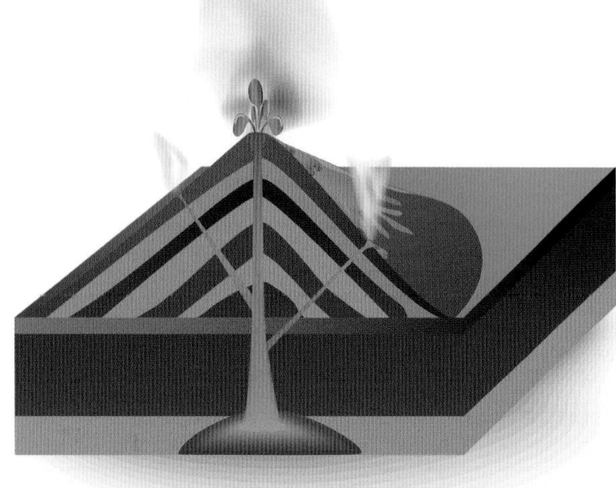

Stratovolcano

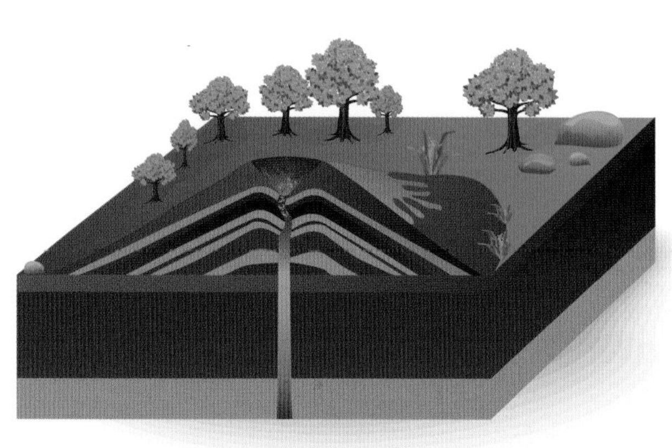

Lava dome

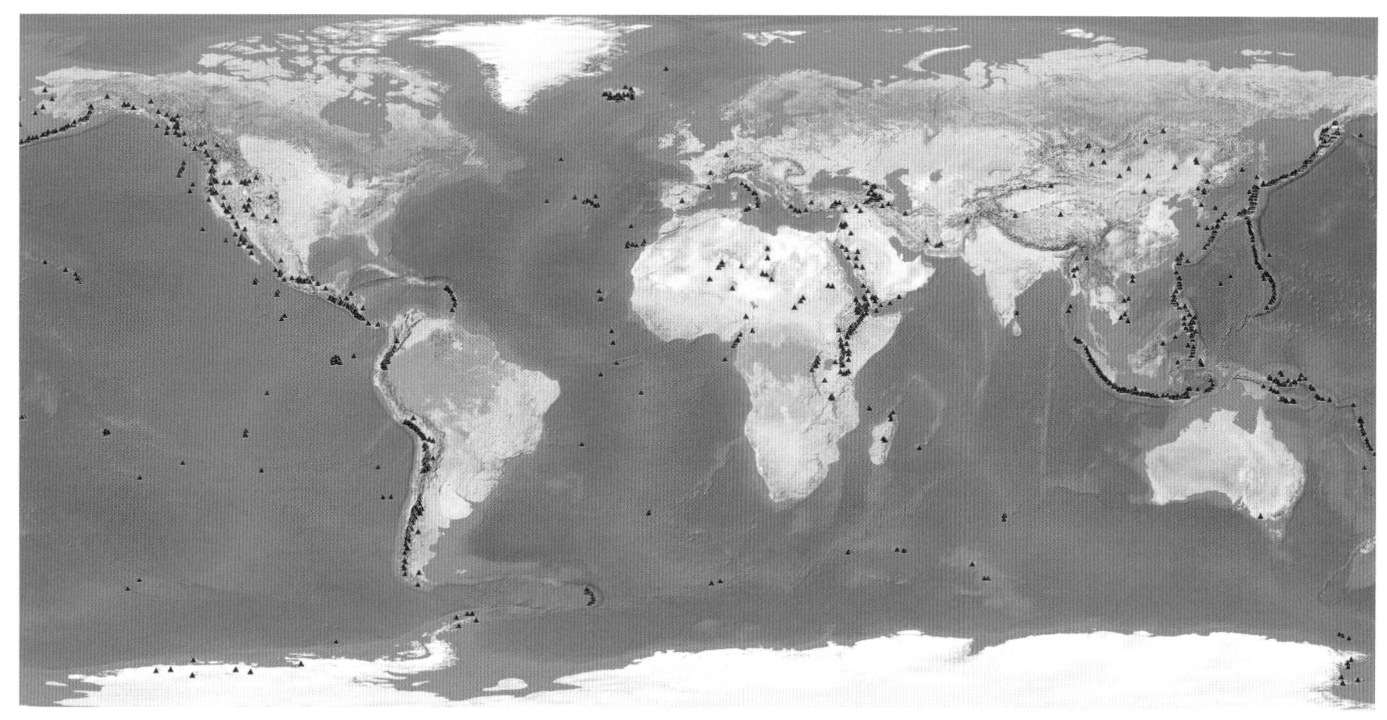

Above *Most of the world's volcanoes are located along continental plate boundaries. They are particularly common in the so-called Ring of Fire, a horseshoe-shaped belt about 24,855 miles (40,000 kilometers) long and up to about 311 miles (500 kilometers) wide around the Pacific Ocean that contains a nearly continuous series of subduction zones. About 850–1,000 volcanoes that have been active during the past 11,700 years—about two-thirds of the world's total—are found within the Ring of Fire.*

Below *Freshly cooled lava in Volcanoes National Park, Hawaii. This undulating, rope-like lava is known as pahoehoe. A type of basaltic lava, pahoehoe forms surface features such as this as the very fluid lava moves under a thin, solidifying but still plastic surface crust.*

pyroclastic flows (see below). Loose tephra layers also often spawn dangerous lahars—volcanic mudflows.

The material expelled during a volcanic eruption may take the form of volcanic gases (mostly steam, carbon dioxide and sulfur dioxide), lava (magma) or tephra (solid material thrown into the air). Tephra is created when the rapid expansion of hot volcanic gases blows apart the magma inside a volcano. As magma rises and the pressure on it decreases, dissolved gases come out of solution and burst open the magma, causing material to fly from the volcano. Small particles of tephra are called volcanic ash; large pieces, which can measure more than 4 feet (1.2 meters) across and weigh several tons, are called volcanic bombs.

Lava varies in silica content, and hence viscosity. It may be felsic, which erupts as domes or short, stubby flows, often associated with explosive volcanism; andesitic, which is characteristic of stratovolcanoes; mafic, which is usually hotter than felsic lava and occurs in a wide range of settings; or ultramafic, the hottest type, which today is very rare.

Eruptions may be magmatic, phreatomagmatic, or phreatic. Magmatic eruptions are mostly caused by the

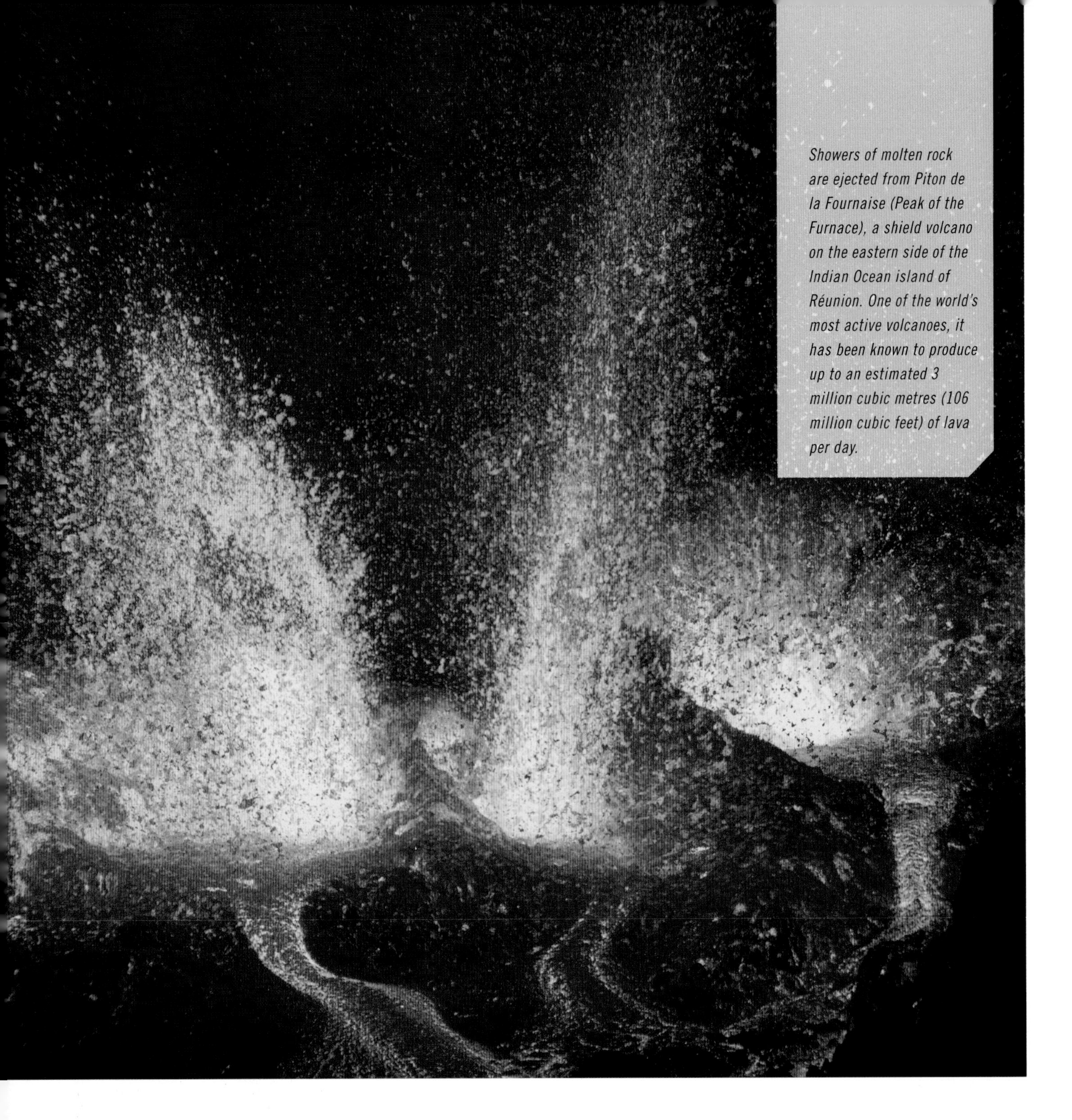

Showers of molten rock are ejected from Piton de la Fournaise (Peak of the Furnace), a shield volcano on the eastern side of the Indian Ocean island of Réunion. One of the world's most active volcanoes, it has been known to produce up to an estimated 3 million cubic metres (106 million cubic feet) of lava per day.

release of gas due to decompression – violently explosive if the magma has high viscosity with much dissolved gas, or relatively gentle if not. Phreatomagmatic eruptions occur when rising magma comes into contact with groundwater, which becomes superheated and creates a rapid build-up of pressure. Phreatic eruptions are also the result of the superheating of groundwater by hot rock or magma, but the erupted material is all existing rock, not new magma.

Eruptions can also be either effusive, featuring less-viscous magma from which gas can easily escape and which flows gently down the slopes; and explosive, in which viscous magma traps gases that build in pressure until they violently break free.

When a volcano undergoes an eruption that produces more than 1,000 cubic kilometres (240 cubic miles) of volcanic deposits during a single explosive event, it's dubbed a supervolcano. Examples include the Yellowstone Caldera in Yellowstone National Park and Ngorongoro Crater in Tanzania.

Large eruptions can send volcanic ash and sulphur dioxide gas high into the atmosphere. There, sulphur dioxide forms sulphuric acid aerosols, which reflect causing what is known as a volcanic winter. The Russian famine of 1601–03, in which around two million people perished, is believed to have been the result of a volcanic winter that was caused by the eruption of Huaynaputina in Peru in 1600.

The release of carbon dioxide by volcanic eruptions has had a significant impact on global climate in the past and may have caused mass extinctions, including the devastating Permian extinction in which some 90 percent of species were wiped out.

Eruptions can aso bring benefits. For example, volcanic ash and weathered lava produce fertile, nutrient-rich soil.

PYROCLASTIC FLOWS

Sometimes, magma and rock broken up by explosive volcanic activity mixes with volcanic gases to form hot (up to 1,560°F/ 850°C), often incandescent mixtures that sweep along close to the ground at velocities of up to 450mph (725 km/h). Such a pyroclastic flow during the 1902 eruption of Mount Pelée on Martinique destroyed the coastal city of Saint-Pierre, killing nearly 30,000 people. Pyroclastic flows are also believed to have been responsible for the destruction of Pompeii in Italy in 79 CE.

Steam rises from the crater of Mount Bromo (front left) an active volcano on the Tengger Massif in East Java, Indonesia. To the volcano's right is Mount Batok, an inactive cinder cone.

// Volcanic and geothermal features

The presence of hot magma close to the Earth's surface creates numerous different landscape features, particularly when there are also large quantities of groundwater present.

VOLCANIC FEATURES

Caldera: when a large explosive eruption takes place, the subterranean chamber in which magma is stored may be emptied, causing the mountain above to collapse into a crater known as a caldera. Calderas can also form when an explosive eruption removes the summit of a stratovolcano. Often circular and steep-sided, they may be more than 25 kilometres (15 miles) across and several kilometres deep.

Cinder cone: cinder-like material ejected from a vent can build up to form a 30–400-metre-tall (100–1,300 foot) cone-shaped hill. Such eruptions tend to be short-lived.

Fissure vent: a flat, linear fracture through which lava emerges.

Lava flow: lava that streams out of a volcano – particularly during effusive eruptions – moves downhill under gravity until it cools and solidifies. Lava can travel at speeds approaching 50 km/h (30 mph) and for more than 100 kilometres (60 miles). Its chemistry and viscosity and the volcano's eruption type all affect the appearance of the flow. Mafic lava may be a´a, characterized by a rough surface and associated with cooler basalt lava flow, or pahoehoe, which has a smoother, ropey or wrinkly surface and is associated with more fluid lava. When submarine volcanoes erupt, seawater quickly cools the emerging lava, creating so-called pillow lava.

Lava tube: when lava is channelled into a single stream whose outer surface solidifies, it forms a tube through which liquid lava continues to flow. If all the lava eventually flows out, it leaves a linear cave or lava tube.

Lava dome: formed when viscous magma erupting effusively onto the surface piles up around the vent, a lava dome can be several hundred metres high and thousands of metres across. Typically, a small dome expands as more magma is forced up into its interior.

Cryptodome: sometimes viscous lava is forced upward but doesn't erupt, causing a bulge on the surface of a volcano known as a cryptodome.

GEOTHERMAL FEATURES

Fumarole: an opening in the Earth's surface that emits steam and volcanic gases such as sulphur dioxide and carbon dioxide. The steam is created when magma comes into contact with groundwater; the gas is usually emitted directly from magma. Fumaroles, often with surface deposits of sulphur-rich minerals around the vents, can form clusters or large fields numbering in the thousands; they may be active for centuries or disappear within weeks. If the water table is close to the surface, fumaroles can become hot springs.

Hot spring: when groundwater comes into contact with hot rock it may emerge as a thermal or hot spring. In volcanic regions, the source of heat is usually magma, and the water may be close to boiling point.

A miner collects chunks of sulphur in the crater of the active volcano Kawah Ijen in East Java, Indonesia. The miners use pipes to bring sulphurous gases from inside the volcano to the surface. As the gases cool, they condense into a liquid and eventually solidify.

Geyser: a rare type of hot spring in which hot water and steam are periodically ejected as a powerful jet. Geysers erupt when water at the base of a tubular opening in the rock becomes superheated after coming into contact with magma but doesn't boil because it's under intense pressure. As some of the water is pushed out of the tube, the pressure decreases and some of the remaining water suddenly flashes into steam and expands, pushing the water above through the tube into the air. Groundwater then flows back into the tube, recharging the geyser. Hence the eruption cycle tends to repeat regularly, with recharge times that can be a few minutes or as long as a few days. Old Faithful in Yellowstone National Park has erupted every 60–90 minutes for hundreds of years, sending as much as 7,000 gallons (32,000 liters) of boiling-hot water more than 165 feet (50 meters) into the air.

Mud pot: in geothermal areas where water is in short supply, hot springs and fumaroles may be replaced by pools of bubbling viscous, acidic mud known as mud pots. Particularly colorful examples are sometimes referred to as paint pots.

Mud volcano: these eruptions of a slurry of fine-grained minerals, water and gas (typically methane) are not necessarily geothermal. Although some form over the vent of an igneous volcano, most aren't associated with magmatic activity. Instead, the mixture is forced to the surface via faults and fissures by pressure caused by tectonic forces, such as those in subduction zones, or overlying sediments. Mud volcanoes may be as much as 6 miles (10 kilometers) across and 2,300 feet (700 meters) high. Like igneous volcanoes, they can form both on land and under the sea. The temperature of the erupted material is much lower than that in igneous volcanoes.

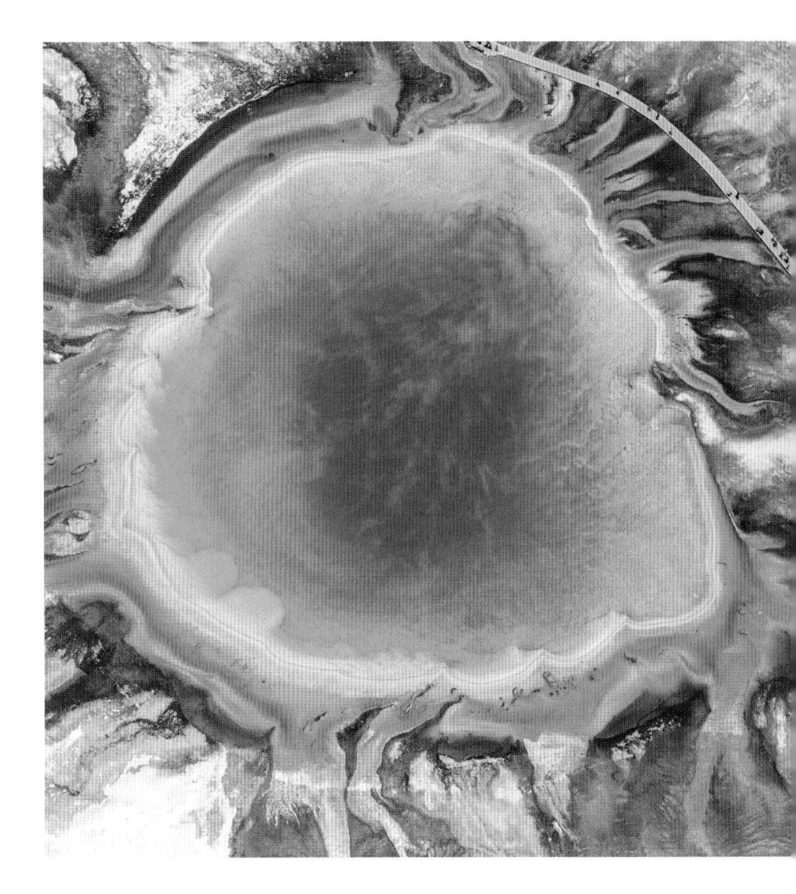

Top right *The Grand Prismatic Spring in Yellowstone National Park, Wyoming, USA, is the third-largest hot spring in the world. About 361 feet (110 meters) in diameter and 164 feet (50 meters) deep, the spring discharges around 462 gallons (2,100 liters) of 158°F (70°C) water per minute. Mats of cyanobacteria are responsible for the concentric rings of color around the pool—as the water spreads out from the center and cools, it creates a temperature gradient, with different bacteria species inhabiting different temperature zones.*

Right *Castle Geyser in Yellowstone National Park, Wyoming, USA. The geyser's 1,000-year-old cone is made from a substance known as sinter, which in this case consists of opaline silica. It erupts every 9–11 hours, sending hot water 98 feet (30 meters) into the air for 20 minutes before changing to a noisy steam phase that lasts about 40 minutes.*

// Earthquakes

Among the deadliest natural disasters, earthquakes cause the Earth's surface to shake as pent-up seismic energy is suddenly released.

Earthquakes typically occur when two pieces of tectonic plate suddenly slip past one another along a discontinuity known as a fault. "Normal" faults involve plates pulling up or down away from each other; at reverse or thrust faults, plates press closer together so that one moves under another; and strike-slip or transform faults feature plates that are moving horizontally parallel to each other in opposite directions.

Plate edges are rarely smooth and they often lock together. As the rest of the plate continues to move, stress builds up at the jammed point until the fault ruptures, suddenly and violently releasing energy into the surrounding rock in the form of seismic waves that radiate outwards from the initial point of rupture, known as the hypocenter or focus, like ripples on a pond. As these waves pass through rock, they make it shake, often rupturing the land surface and nearby infrastructure.

An earthquake's power or magnitude depends on the size of the fault and the amount of slip. The best-known measure for earthquake magnitude is the Richter scale, developed by Charles F. Richter in 1935, although this is gradually being replaced by other scales. All use logarithmic values: every unit increase in magnitude represents a 10-fold increase in the amplitude of the ground shaking and a 32-fold difference in the amount of energy released. An 8.6-magnitude earthquake releases roughly 10,000 times as much energy as the atomic bomb dropped on Nagasaki, Japan in World War II.

Many earthquakes are so weak that they cannot be felt; others are so violent they devastate entire cities. Estimates suggest that around 500,000 earthquakes occur each year, about a fifth of them strong enough to be felt and about 100 large enough to cause serious damage.

The most powerful quakes occur along thrust or reverse faults on convergent plate boundaries. Some 90 percent of all quakes and 81 percent of the largest occur offshore around the rim of the Pacific Plate—the "Ring of Fire"—so most of the largest occur deep below the ocean surface.

The most powerful earthquake ever measured on a seismograph took place on May 22, 1960. Reaching a magnitude of 9.5, its epicenter (the point at the surface directly above the hypocenter) was near Cañete, Chile.

Those with a magnitude of 8 or more are known as megathrust earthquakes. They are responsible for about 90 percent of the total seismic energy released worldwide.

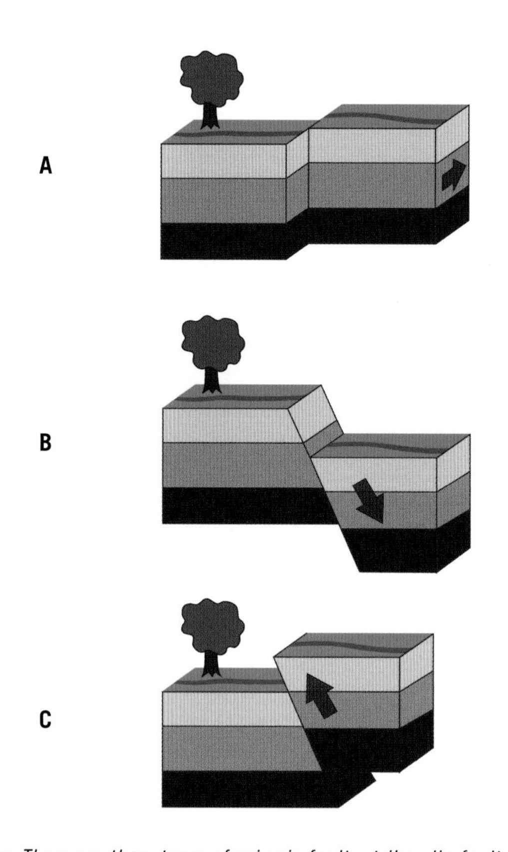

Above *There are three types of seismic fault: strike-slip faults (A), where rocks are sliding past each other horizontally with little or no vertical movement; normal faults (B), where two blocks of crust are pulling apart; and reverse faults (C), where one block of crust is sliding over another.*

Earthquakes associated with strike-slip faults typically have magnitudes of less than 8, while those on normal faults generally have magnitudes of less than 7.

Earthquakes can also be triggered by volcanic activity, thus potentially serving as an early warning of an impending eruption, or man-made, caused by mine blasts and nuclear tests or even the movement of water in a reservoir.

Some earthquakes are preceded by a number of foreshocks. These are followed by the "mainshock"—the large, primary earthquake—and then often multiple aftershocks (tremors with smaller magnitudes as the crust around the displaced fault plane adjusts). Aftershocks may continue for years. Sometimes a region is struck by an earthquake swarm—a series of earthquakes of similar magnitude, none of them obviously the mainshock, over a short period of time.

Secondary effects of earthquakes include landslides and avalanches, or floods when debris dislodged by a quake

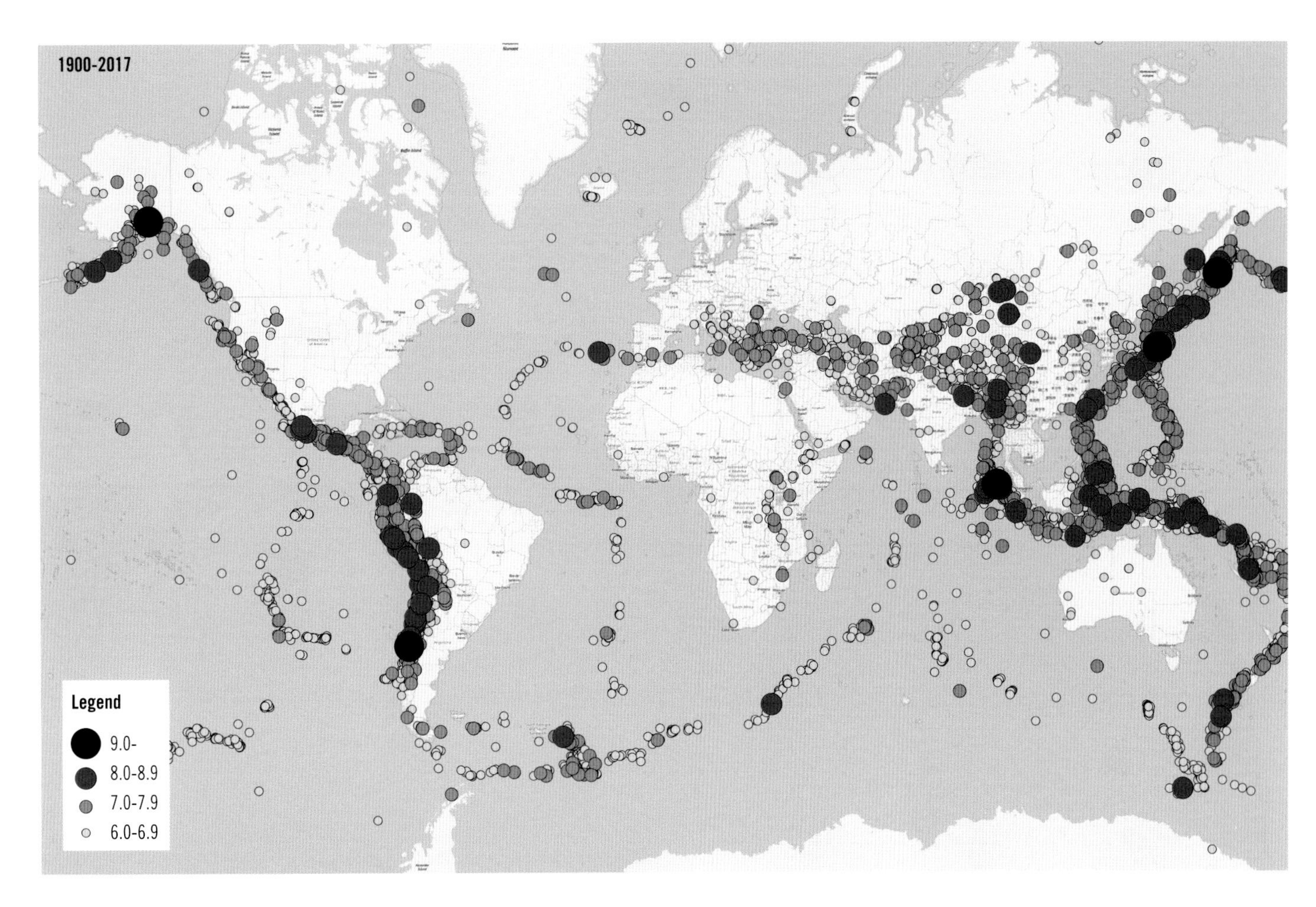

Above *Just after midnight on November 14, 2016, the 7.8-magnitude Kaikoura earthquake struck the South Island of New Zealand. Ruptures occurred on as many as 25 different faults—the most recorded for a single quake—causing widespread damage, including significant buckling of State Highway 1 as seen here.*

Above *Map showing the locations of earthquakes around the world between 1900 and 2017. As is true of volcanoes, most earthquakes occur along plate boundaries and they reach their highest concentrations and highest magnitudes around the Pacific Ring of Fire.*

dams a river or tremors cause a dam to burst. By damaging electrical-power lines and rupturing gas pipes, quakes can trigger fires that are more destructive than the earthquake itself. Earthquake tremors can also cause water-saturated soil to temporarily transform from solid to liquid, a process called soil liquefaction that can make overlying structures such as buildings and bridges tilt, slide away or sink into the ground. Probably the most devastating secondary effects of all are tsunamis, produced when earthquakes in the open ocean cause a large volume of water to be abruptly moved up or down (see page 120).

Earthquakes can result in large-scale loss of life, typically due to the collapse of buildings and other structures or by generating tsunamis. More than 830,000 people died when an earthquake struck China's Shaanxi province on January 23, 1556, while almost a quarter of a million people were killed by the Indian Ocean tsunami generated by an earthquake off Sumatra, Indonesia, on December 26, 2004.

// Mountains

Massive landforms that rise high above the surrounding landscape, mountains alter local climate and weather, and often define national borders.

In general, a landform that rises more than 1,000 feet (300 meters) above its surrounding area, or more than 2,000 feet (600 meters) above sea level, is considered to be a mountain, while anything lower is a hill. Mountains also tend to be steeper than hills. High areas with no specific peak are known as plateaus.

Although some mountains exist as isolated summits, most are found in ranges. A collection of linked mountain ranges is termed a mountain belt. About a quarter of the Earth's landmass is mountainous.

Mountains are formed by three main processes: volcanism, thermal expansion of the crust and crustal shortening due to tectonic activity (where parts of the crust are either destroyed or buckled in a subduction zone).

Most volcanic mountains form due to the build-up of lava on the surface. However, magma can also push the crust up from below and then harden before it erupts, forming what are known as dome mountains. Volcanic mountains typically form along convergent plate boundaries, but can also emerge far from plate boundaries when the crust moves over a hotspot in the mantle.

Thermal processes account for much of the high terrain in Africa and Antarctica. Like most materials, rock expands as it is heated. If the lithosphere of a region is unusually hot, its expansion can raise the surface level to form a mountain range or plateau.

The majority of mountains, however, are formed by tectonic activity. The world's tallest mountain ranges are made up of fold mountains that formed when two tectonic plates collided, buckling the crust. Sometimes folding leads to the formation of parallel bands of rock: less resistant rocks weather away more quickly, creating valleys, while tougher crystalline rocks remain, forming high peaks. The Himalaya, for example, are the remains of crystalline rocks that solidified deep in the crust and were then thrust to the surface by folding. Two-thirds of the world's 40 highest mountains are in the Himalaya, including Mount Everest, whose summit, at 29,032 feet (8,849 metres), is the highest point on Earth.

In other examples of folding, when plates collide and the lighter, more buoyant plate rides up over the denser plate, it scrapes off the leading edge of the descending plate. Eventually, several slices of continental material are stacked up on the overlying plate. As the crust plows forward, it buckles and folds the scraped-off material into mountains.

Tectonic activity can also produce block mountains or horsts, formed when material on one side of a tectonic fault is raised up relative to the other. The lower block of crust is known as a graben (or half graben, depending on its configuration). Some block mountains consist of an uplifted block between two faults, while others are tilted blocks, raised on one side and/or subsiding on the other. Block-faulted ranges commonly form within continents, far from collision or subduction zones. At the edges of such ranges, sedimentary rocks are often tilted by 90°. If these rocks are resistant to erosion, they can form narrow, sharp-crested ridges called hogbacks that run parallel to the range.

Because most mountain ranges formed where two plates converged, and the convergence zones intersect with the boundaries of other plates, many of the world's mountain ranges link together to form two very long mountain systems. The Circum-Pacific System surrounds most of the Pacific Ocean basin, while the Alpine-Himalayan System runs from Morocco, through Europe, across Turkey and Iran,

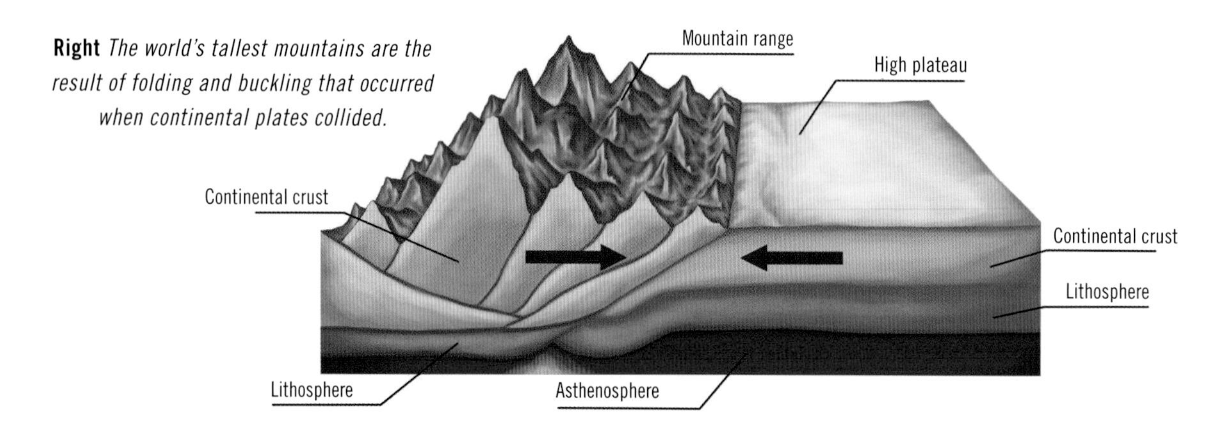

Right *The world's tallest mountains are the result of folding and buckling that occurred when continental plates collided.*

Above *The Himalaya contain many of Earth's highest peaks, including the highest, Mount Everest, and more than fifty mountains with heights of more than 23,500 feet (7,200 meters). The 1,500-mile (2,400-kilometer) range was created by the subduction of the Indian tectonic plate under the Eurasian Plate.*

and through the Himalaya to Southeast Asia. Mountains that are not within these systems are mostly residual, formed during ancient continental collisions hundreds of millions of years ago or by non-tectonic processes.

Climate in mountains becomes colder with increasing elevation at a rate of about 28°F per mile (9.8°C per kilometer) of altitude. The temperature gradients associated with mountains affect ecosystems: taller mountains often have ecological bands populated by communities adapted to a particular climatic range, a phenomenon known as altitudinal zonation.

Tall mountains can significantly influence weather patterns by interrupting the flow of air, often leading to what are known as rain shadows, where the landward flanks of coastal mountains are drier than those facing the sea. Water flowing from mountains feeds most of the world's rivers; more than half of humanity depends on mountains for their water.

Above *Mountains can be classified according to how they were created.*

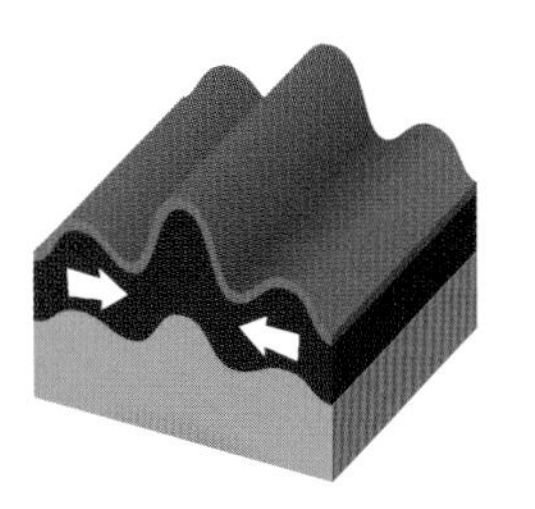

Fold mountains
Formed when continental plates collide.

Upwarped mountains
Formed when molten rock rises up and pushes layers of rock up.

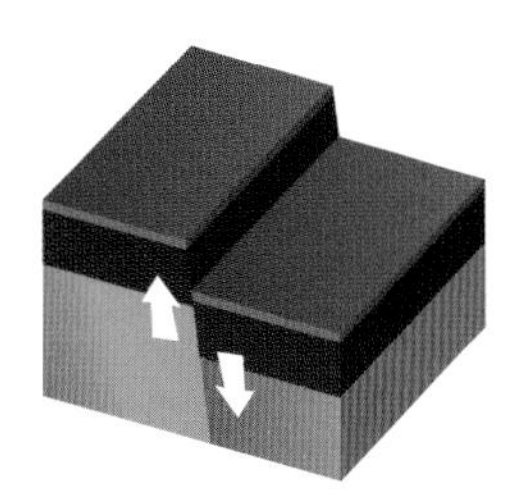

Fault-block mountains
Formed when a block of rock drops down compared to other blocks.

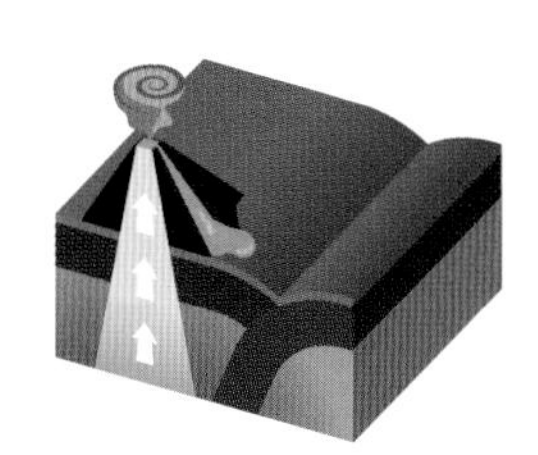

Volcanic mountains
Formed when magma spills out and hardens on the Earth's crust.

// Islands

From barren rock outcrops to vast "mini continents," islands take many different forms.

Left *The British Isles formed part of mainland Europe before rising sea levels flooded the English Channel.*

Islands are bodies of land entirely surrounded by water—be it ocean, sea, lake, or river. Very small islands are called islets, holms, or skerries. Those in rivers are aits or eyots; sedimentary islands in the Ganges delta are known as chars. A group of geographically or geologically related islands is an archipelago.

There are two main types of island: continental and oceanic. The former lie just off a mainland on the continental shelf (see page 94), while the latter may be located far out in open ocean basins. Larger islands tend to be continental.

Many continental islands formed relatively recently as melting ice sheets caused sea levels to rise at the beginning of the current interglacial period. The British Isles, for example, were part of mainland Europe before the English Channel flooded. So-called microcontinental islands form when a piece of continental crust becomes separated from the mainland. Some continental islands are much older, having come into being when the supercontinent of Pangaea broke apart (see page 16). Others formed as erosion and weathering severed a link with the adjacent mainland.

Most oceanic islands are volcanic in origin (the exceptions are those formed when tectonic forces lift the ocean floor above sea level). When an undersea volcano erupts, lava accumulates on the sea floor, slowly building until it eventually rises above the ocean surface to create an island. In Hawaii, such piles of lava rise as much as 31,800 feet (9,700 meters) above the sea floor. Volcanic oceanic islands often form over mantle hotspots. As an oceanic plate moves over a hotspot, an island chain may be created. Over time, the older islands eventually sink beneath the surface once more, becoming guyots (see page 98). Oceanic islands can also form at plate boundaries: in rift zones, where diverging tectonic plates pull apart and magma rushes up to fill the gap; and in subduction zones, where one tectonic plate is sliding under another, throwing up an island arc.

These oceanic islands may have limited biological communities, mostly plants, seabirds and insects. Although vegetation is often abundant, the diversity of plant species will be low. Isolation can lead to unusual forms evolving, with both gigantism and dwarfism common on islands.

Barrier islands are long, narrow islands that lie parallel to coastlines, usually separated from the mainland by a quiet lagoon. Most barrier islands form as ocean currents slow down and deposit sand and other sediments on the sea floor. Eventually, this sandbar rises above the surface and becomes an island. These sorts of barrier islands tend to be quite dynamic, changing shape and even disappearing

ISLAND VERSUS CONTINENT

There is no established definition of what constitutes a continent, and what distinguishes islands from continents. Greenland, with an area of almost 850,000 square miles (2.2 million square kilometers), is generally considered the world's largest island, while Australia, which has an area of 2.9 million square miles (7.7 million square kilometers), is seen as the smallest continent. Arguments supporting Australia's continent status include the fact that it is the largest landmass on its continental plate and has its own distinct flora, fauna and culture. However, none of these factors are conclusive.

completely as currents shift and storms occur. Some barrier islands formed as rising sea levels at the end of the last glacial period drowned coastal sand dunes.

Tidal islands are continental islands that are connected to the mainland by a land bridge that is exposed at low tide but submerged at high tide.

Coral islands are low, exposed coral reefs. Organic and inorganic material may accumulate on the coral to form a more substantial but still small, low, sandy outcrop, known as a cay or key. Atolls are a specific type of coral island formed when a reef grows around the perimeter of an oceanic island that then sinks into the sea, leaving a ring of coral around a central lagoon. An atoll can take up to 30 million years to form.

The size and, in some cases, the very existence of islands is dependent on sea level. Falling sea levels expose more land, while rising sea levels can cause islands to shrink or disappear completely. Today, rising sea levels represent an existential risk to many low-lying coral islands and the communities they support, exposing them to increased damage and erosion from storms, and eventually threatening to submerge them.

An underwater volcano rises above the surface of the ocean.

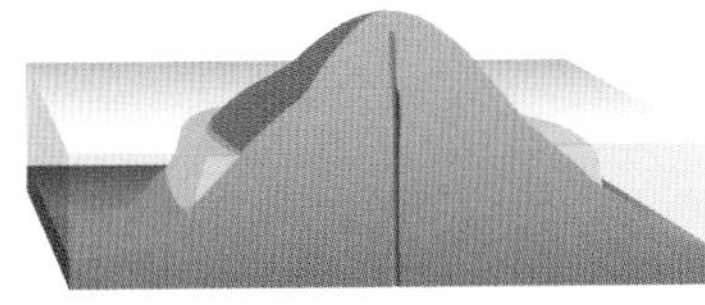

A fringing coral reef forms around the volcanic island.

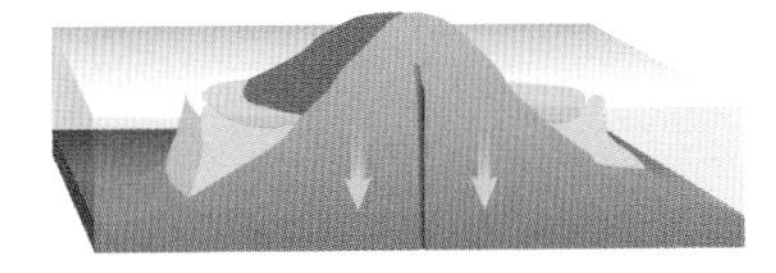

As the volcano subsides, the reef remains and a lagoon forms.

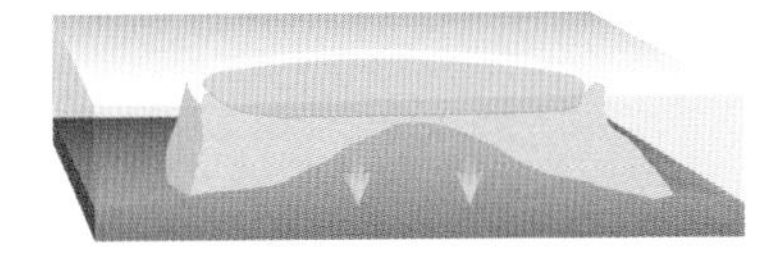

Eventually the volcano is submerged, leaving behind an atoll and lagoon.

Above *The formation of an atoll begins with the growth of a fringing reef around a volcanic island. Eventually, the central island subsides, leaving behind a central lagoon surrounded by the low sandy coral atoll.*

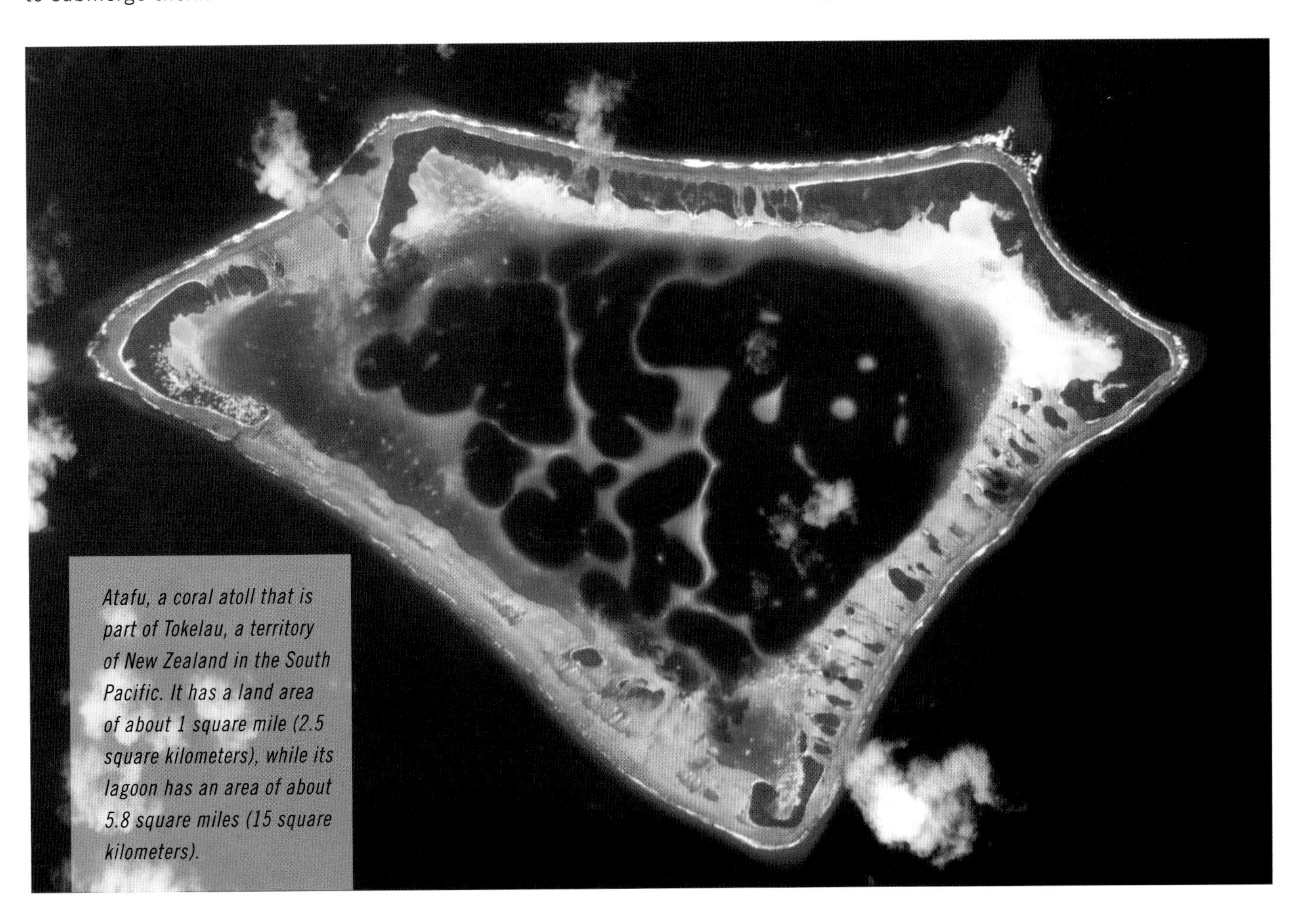

Atafu, a coral atoll that is part of Tokelau, a territory of New Zealand in the South Pacific. It has a land area of about 1 square mile (2.5 square kilometers), while its lagoon has an area of about 5.8 square miles (15 square kilometers).

// Rivers

Naturally flowing watercourses, rivers are powerful, dynamic landscape features, cutting down into bedrock and regularly altering their course.

Small rivers are known variously as streams, creeks, brooks, rivulets and rills, but none of these terms, including river, are strictly defined. Together, rivers and streams cover around 0.1 percent of the Earth's surface.

Rivers flow downhill under the influence of gravity. They begin at a source (or, more often, several sources), often a spring but also potentially a glacier or an area of melting snow, and end at a mouth (or mouths), usually flowing into the sea.

A river's path is known as its course. In their upper courses, rivers typically flow in narrow channels through steep-sided, V-shaped valleys and gorges, forming tumultuous rapids as they tumble over shallow sections and waterfalls as they drop over precipices. The steep gradient makes the water flow quickly, causing it to cut deep into the underlying rock and soil. Where the rock is particularly hard, the water will often change direction, creating what are known as interlocking spurs, a series of ridges that project out on alternate sides of a valley like the teeth of a zip.

In their middle courses, where the gradient is less steep, rivers usually flow through valleys that are wider and shallower. Their channel begins to widen and deepen. As the river differentially erodes its banks, it forms large bends that eventually become horseshoe-like loops called meanders, which gradually migrate downstream. Meanders form as water erodes and undercuts the river bank on the outside of the bend while simultaneously depositing sediment on the inside of the bend, where the river's flow is slower. When a meander becomes cut off from the main river it forms an oxbow lake.

By the time they reach the end of their journey, rivers will typically be flowing through a wide, flat-bottomed valley and across a floodplain—the area around a river that is covered when the river floods. Their channels will be wide and deep.

As they flow, rivers erode away underlying material and carry it downstream. The combined suspended sediment load of the Ganges-Brahmaputra river system is about 2 billion tons (1.8 billion tonnes) a year—the world's highest.

Gradually the gradient becomes even flatter, the speed of flow decreases and much of the sediment carried drops out

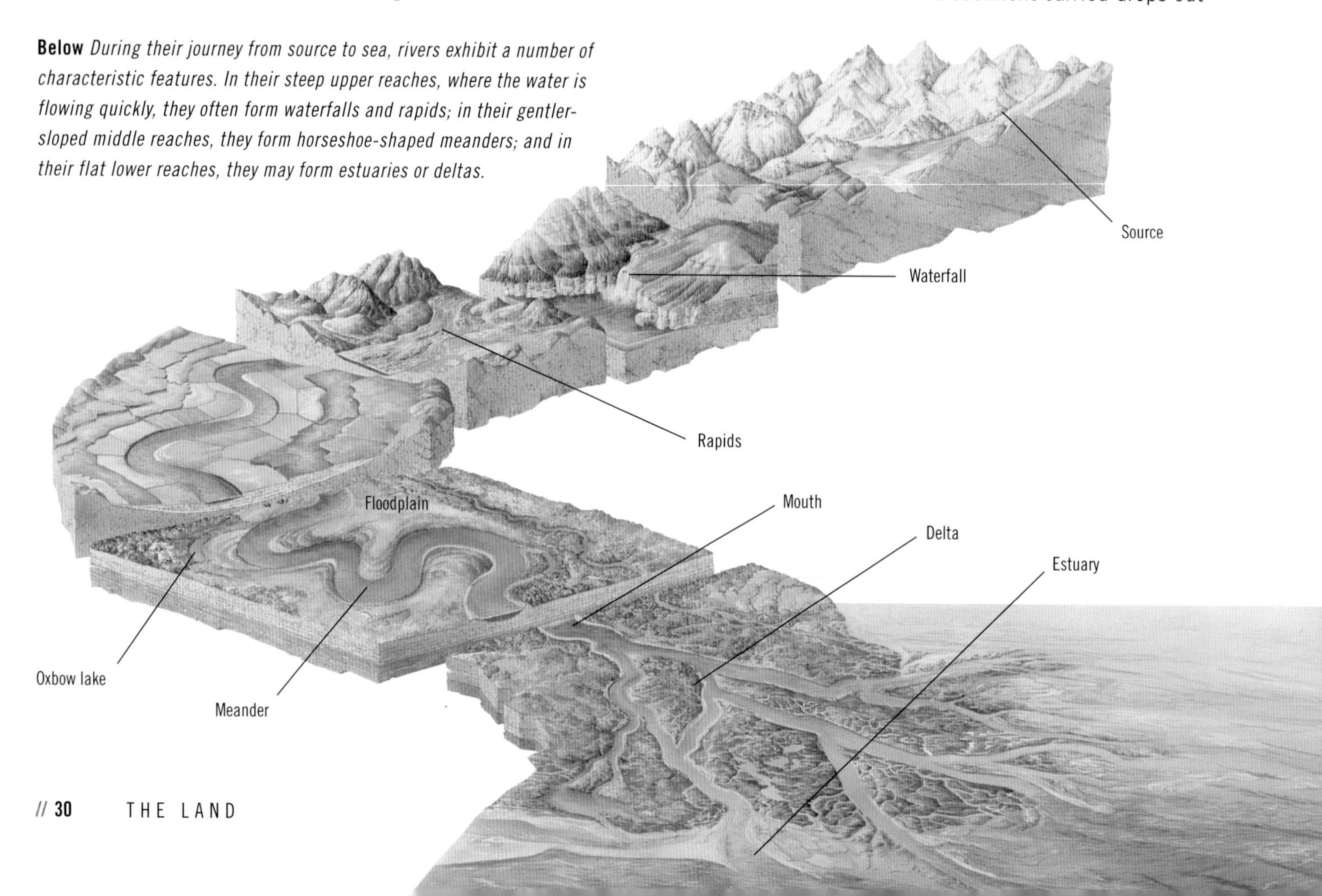

Below *During their journey from source to sea, rivers exhibit a number of characteristic features. In their steep upper reaches, where the water is flowing quickly, they often form waterfalls and rapids; in their gentler-sloped middle reaches, they form horseshoe-shaped meanders; and in their flat lower reaches, they may form estuaries or deltas.*

Above *The Amazon has the largest drainage basin (about 2.7 million square miles/7 million square kilometers) and the largest discharge volume (7,380,000 cubic feet/ 209,000 cubic metres per second or 1,581 cubic miles/6,591 cubic kilometers per year) of any river on Earth.*

of suspension, often leading to the formation of an ever-growing delta splaying out into multiple branches at the river's mouth (see page 36). If the sediment is washed away before it can build up, the river mouth will form a single-channeled brackish estuary (see page 38).

During its journey from source to mouth, a river is fed by smaller streams and rivers, known as tributaries. The total area over which precipitation feeds into a river is its drainage basin or catchment area. The river with the largest average discharge is the 4,000-mile (6,400-kilometer) Amazon, which discharges roughly 7,380,000 cubic feet (209,000 cubic meters) of water into the Atlantic Ocean each second. It also has the largest drainage basin, covering an area of some 2.7 million square miles (7 million square kilometers).

The total volume of water transported downstream is often a combination of surface water flow and a substantial flow through underlying sub-surface rocks and gravels, known as the hyporheic zone. For rivers flowing through large valleys, the flow in the hyporheic zone may be much greater than the visible flow. In karst areas (see page 64), rivers may be almost completely subterranean, flowing underground through caves and caverns.

Most rivers can be classified as either alluvial, bedrock or a mixture of the two. Alluvial rivers flow through channels formed in sediments; bedrock rivers flow through channels cut into the underlying rock.

Rivers with steep gradients, high sediment loads and regularly changing discharge levels may split into multiple channels that continually split and rejoin. These so-called braided rivers may occupy an entire valley floor.

A braided river in Iceland. Consisting of a network of interconnected channels separated by small, often temporary, islands, braided rivers often have high sediment loads.

// Canyons

Deep, narrow, steep-walled valleys, canyons are testament to water's erosional power.

Canyons are known by various names around the world, including gorge, ravine, and defile; gorges are generally considered to be steeper and narrower than canyons. A river that flows through a canyon is called an entrenched river because it doesn't meander and change its course.

Such valleys are typically formed by a combination of erosion and weathering. As streams and rivers flow, they erode away the top layer of their bed, slowly cutting down through soft rock. Harder rock strata remain as cliffs on either side. During especially cold periods, water that has seeped into cracks in the rock walls of the nascent canyon freezes and expands, widening the cracks and eventually causing flakes and chunks of rock to fall away, a mechanism known as frost wedging. Heavy rainstorms can hasten this process, ripping away any loose rocks.

Geological uplift caused by tectonic activity sometimes plays a role in canyon creation. When uplift raises land to form a plateau, any streams or rivers flowing across it will descend as waterfalls at the edge. As the water flowing over the falls erodes the rock layers beneath, the position of the waterfall will move upstream, leaving behind a canyon. The Grand Canyon in Arizona was formed in this way. The largest canyon in the USA, with a volume of almost 148 trillion cubic feet (4.2 trillion cubic metres), it was formed over millions of years after uplift created the Colorado Plateau. It's currently being eroded at a rate of 1 foot (0.3 metres) every 200 years.

In karst regions (see page 64), canyons sometimes form when cave systems collapse. Many of the numerous canyons that are found in the Yorkshire Dales in northern England were created in this way.

Canyons are generally more common in arid areas because erosion and weathering tend to be more localized there. Swift-moving streams and rivers fed by snowmelt in far-off mountains cut down through the rock, while surrounding areas remain intact. They also commonly form in a river's upper reaches, where currents are usually stronger and faster, hastening the rate of erosion.

Very narrow canyons, often with smooth walls, are known as slot canyons. They can be less than 3 feet (1 meter) wide but hundreds of feet deep. They typically form when periodic bursts of rushing water cut into plateaus of soft rock such as sandstone.

Small canyons that are bounded on three sides at their head are known as box canyons. These often form where a spring emerges at the base of a cliff. Water from the spring soaks through the rock until it reaches an impermeable layer, then spreads sideways, weakening the cliff, which collapses to form a canyon.

Some of the world's deepest canyons, with depths

Sunset at Horseshoe Bend, Arizona, USA. Following a period of uplift between five and six million years ago, the Colorado River – which was at that time a meandering river with an almost level floodplain – began to rapidly cut down through the underlying rock, eventually forming Horseshoe Bend and the nearby Grand Canyon.

of more than 5 kilometres (3 miles), are found in the Himalaya. The world's longest known canyon is actually hidden beneath the ice in Greenland. Also called the Grand Canyon, it is 750 kilometres (470 miles) long.

Steep-sided valleys can also be found cutting down through the sea floor of the continental shelf and slope. Such submarine canyons can either be the continuation of river canyons or created by sediment-carrying ocean currents and underwater landslides. Some of these canyons are of a similar scale to the USA's Grand Canyon.

// Waterfalls

When rivers and streams flow over a vertical drop, they form waterfalls.

Waterfalls tend to be most common in the upper and middle reaches of a river or stream and typically occur where the water flows from hard to soft rock. When the softer rock erodes away, only a hard ledge remains and it is over this that the water falls. As a river or stream approaches a waterfall, its velocity increases, increasing the rate at which erosion takes place. The falling water and sediment then erode the plunge pool (the basin at the base of a waterfall into which water flows) below.

Tectonic activity can also create waterfalls. Movement along faults can lead to the formation of a vertical drop as land on one side of the fault sinks or is raised up. Similarly, tectonic uplift can raise an area of land to create a high plateau. Any rivers or streams that flow over these areas will form waterfalls as they leave the higher ground.

Glacial action can also create waterfalls, gouging out deep valleys that leave tributaries, known as hanging valleys, high up on the steep valley sides. Landslides and lava flows that disrupt stream and river beds, and cave formation and collapse in karst areas, can lead to the creation of waterfalls as well.

Waterfalls are dynamic landscape features. Constant erosion by moving water and the sediment it contains can cause a waterfall to "recede" or retreat. As the rock behind the waterfall is worn away, it develops a cave-like structure called a rock shelter. The rocky ledge above, known as an outcropping, eventually collapses, causing the waterfall to move upstream, where the process begins again. The rate of retreat can be as great as 5 feet (1.5 meters) per year and will often lead to the creation of a steep-sided gorge on the downstream side. Niagara Falls, on the border of Canada and the USA, has retreated 7 miles (11 kilometers) from its original position.

Waterfalls can take many forms, ranging from the classic single stream falling over a cliff edge, known as a plunge waterfall, to cascades that descend over a series of rock steps, block waterfalls that descend in a wide sheet, or horsetail waterfalls, which maintain contact with the rock that underlies them.

Angel Falls, the world's tallest waterfall, is so high (979 metres/3,212 feet) that the water that falls over it often doesn't reach the river below: when the air pressure is stronger than the water pressure of the falls, the water becomes mist and is blown away. The cloud formed by the mist is called cataractagenitus.

The waterfall with the largest flow is Khone Falls on the Mekong River in Laos, over which an estimated 409,650 cubic feet (11,600 cubic meters) of water flows each second.

Below *The evolution of a waterfall. Water flowing over the lip of a layer of harder rock causes a plunge pool to form at the waterfall's base. Over time, the softer rock behind the waterfall is worn away, creating a cave-like structure known as a rock shelter. The rocky ledge above, known as an outcropping, eventually collapses, causing the waterfall to move upstream, where the process begins again.*

Right *Located on the Guichun River on the border between Vietnam and China, Detian Waterfall has a drop of about 200 feet (61 meters). The waterfall's multiple tiers reflect the different layers of the underlying dolomitic limestone bedrock, which each have a different hardness.*

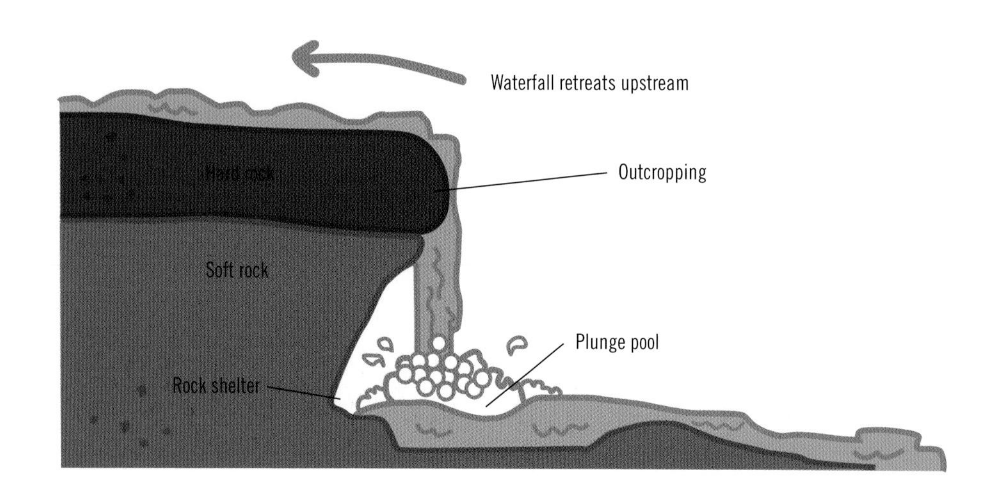

// River deltas

When rivers reach the sea, they drop the load of sediment they're carrying, often leading to the creation of new land in the form of fertile deltas.

The delta of the Avacha River on the Kamchatka Peninsula, Russia. Much of the sediment that flows into the delta and then out into the Pacific Ocean comes from the numerous active volcanoes that are present on the peninsula.

Deltas form at the mouths of large rivers when they deposit sediment more rapidly than the sea removes it. The land around a river's mouth tends to be relatively flat and the banks far apart. As the river water approaches these lower reaches, it slows down greatly, causing most of the sediment it is carrying to drop out of suspension. The heavier, coarser material settles first, followed by the finer material, which is known as silt or alluvium.

If the sediment is not carried away, it grows into what is known as a deltaic lobe. This lobe forces the river to change its course or split into more than one channel, which tends to slow it down even further. As each new channel forms (a process known as avulsion), the delta grows laterally, producing the characteristic triangular shape that led it to be named after the triangle-shaped upper-case Greek letter delta. Within a mature deltaic lobe there will be numerous smaller, shallower channels called distributaries that branch off from the river's primary channel.

Deltas can be divided into three main types according to their shape: arcuate (arc-like) deltas are triangular or fan-shaped (for example, the Nile River delta in Egypt); cuspate deltas, where the land around the river mouth juts out into the sea like an arrow, are formed when waves are strong at the river mouth (for example, the Tiber River delta in Italy); and bird's-foot deltas, where a river has split and each new channel has a delta of its own that juts out into the sea (for example, the Mississippi River delta in the USA). Although ocean deltas are the best known, deltas can also form where a river enters an estuary, lake or even another river.

Deltas can also be classified according to the main control on sediment deposition. In wave-dominated deltas, ocean waves deflect sediment that flows out of the river mouth, pushing it along the coast and possibly causing the delta to retreat. In tide-dominated deltas, which typically form in areas that experience large tidal ranges and often have a dendritic (branching) structure, the flow of water with the tides mediates the deposition of sediment. River-dominated deltas form in areas where wave and tidal current energies are low and hence have little effect on the discharge of water and sediment. Gilbert deltas are steeper, forming where rivers deposit particularly coarse sediment, typically when they empty into lakes. Estuarine deltas form

Above *Delta formation. As a river approaches its mouth, its flow slows down, causing the sediment it is carrying to sink to the river bed. If it isn't then washed away, the sediment builds up in layers that eventually block the main channel, which splits into new branches. As this process is repeated, a delta slowly forms, growing out from the river mouth.*

when a river empties into an estuary rather than directly into the ocean.

The accumulation of nutrient-rich sediment in a river delta creates an extremely fertile environment, and many are cultivated. For this reason, deltas have played an important role in the development of human civilization and tend to support diverse ecosystems.

The tide-dominated delta formed by the Ganges and Brahmaputra rivers (among others) as they flow into the Bay of Bengal is the world's largest, covering an area of more than 40,540 square miles (105,000 square kilometers). Straddling India and Bangladesh, it is also one of the world's most fertile regions; about two thirds of the population of Bangladesh work in agriculture on the delta floodplains or fish within the delta.

On very rare occasions, a river will form an inland delta, where it drops its sediment onto a plain. The best-known example is the wildlife-rich Okavango delta in Botswana. Here, the Okavango River spreads out and evaporates across a flat area of the Kalahari Desert

Where rivers are managed by people, usually through the creation of dams along their length, their deltas can be threatened. In Egypt, for example, the construction of the Aswan Dam during the 1960s reduced annual flooding of the Nile delta, causing it to shrink, the erosion caused by waves in the Mediterranean Sea removing sediment faster than the Nile could replace it.

Whether or not a river forms a delta is also determined by the size and shape of the continental shelf at its mouth. If the shelf is particularly narrow or there is a large canyon carved into it, the sediment cannot accumulate and a delta will not form.

Left *The Okavango Delta in Botswana is an example of an inland river delta. Each year, the Okavango River floods, causing the delta to swell to three times its permanent size. All of this water eventually evaporates or is transpired by vegetation and hence doesn't flow into the ocean.*

// Estuaries

When rivers and streams flow freely into the sea, they form estuaries, highly productive water bodies where fresh and salt water mix.

The mouth of a river is effectively a semi-enclosed coastal water body with a free connection to the sea, and as such may also be known as a bay, lagoon, sound, inlet or slough. Subject to tides and to changing levels of input from rivers and streams, an estuary's salinity will vary widely. Because fresh water is less dense than salt water, an it will often be stratified, with a distinct layer of fresh water floating on the surface. Within an estuary, tidal movements cause the water level and the salinity to rise and fall over the short term, while seasonal change has the same effect over longer periods.

Most of the world's estuaries formed when sea levels

A series of estuaries along the coast of Guinea–Bissau, including that of Rio Geba. The discolouration around the rivers' mouths is caused by dissolved organic matter and sediment carried down from the land.

Types of estuary.

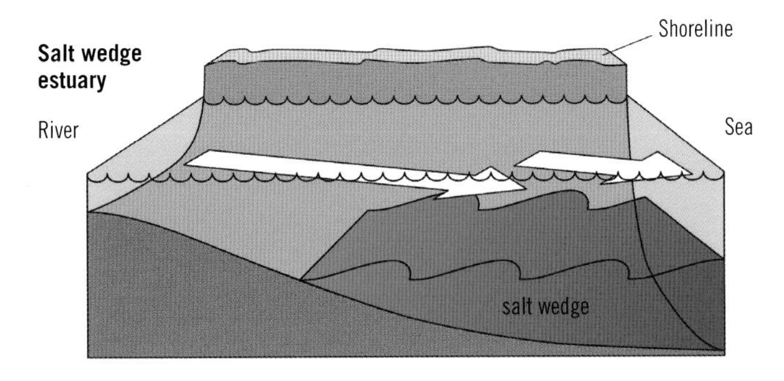

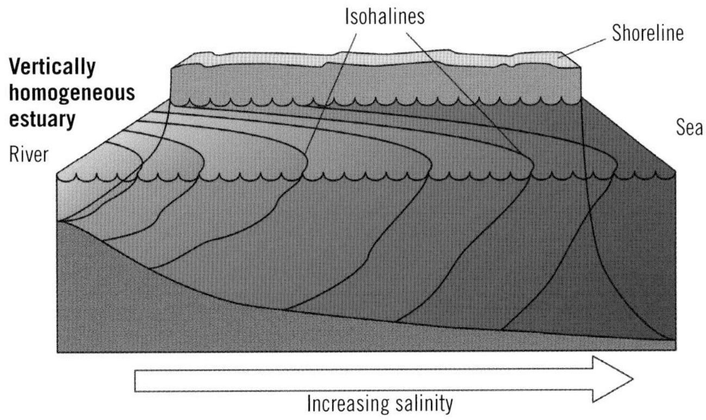

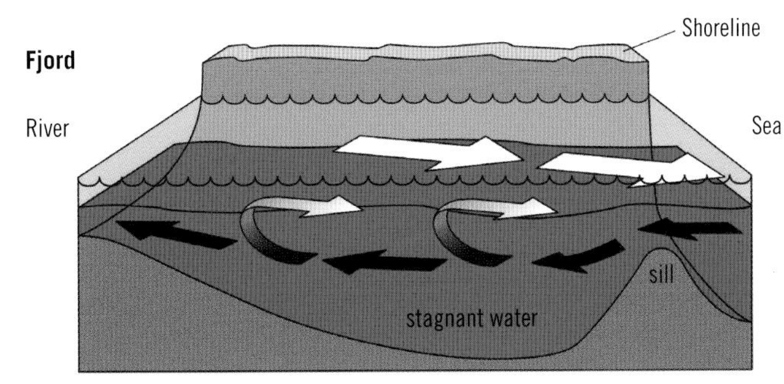

began to rise about 10,000–12,000 years ago as the Earth entered an interglacial period and ice sheets and glaciers began to melt (see page 40).

Estuaries are among the world's most productive ecosystems. They support a diverse assemblage of species, acting as breeding sites and nurseries for many fish species and as stopovers for migratory birds. The mixing of fresh and salt water, as well as the influx of sediment, provides high levels of nutrients, which stimulate primary production. About three quarters of commercial fisheries use estuaries as nursery grounds. The diversity of habitats associated with estuaries, including mangrove forests, marshes, seagrass beds, oyster beds, and mudflats, all contribute to the levels of biodiversity.

Rivers flowing into freshwater lakes also form estuaries; river water tends to be warmer and less dense, and to have different chemical characteristics to lake water. As in coastal estuaries, these areas act as natural filters for runoff and provide nursery grounds for many animal species.

Although they are influenced by the tides, coastal estuaries are generally protected from the full force of ocean waves, winds, and storms by geographical features such as reefs, barrier islands, and sand spits. They also often support marshes, mangroves, and other habitats that act as storm buffers and filter the water flowing into the ocean.

Estuaries are often classified according to how they were created. Four different types are recognized: coastal plain estuaries (also known as rias), created when sea-level rise inundates an existing river valley; tectonic estuaries, formed by land movement related to tectonic activity; bar-built estuaries, which are partially separated from the sea by sand spits and/or barrier islands formed by deposition of river sediment; and fjords, which are steep-sided inlets created by glacial erosion. In temperate regions, coastal plain estuaries are the most common type, while bar-built estuaries are most usual in the tropics.

Offering access to fresh water, a good source of fish and other game, and opportunities for river and maritime trade, estuaries were popular sites for early human settlements, and many of the world's earliest civilizations developed around estuaries. Today, roughly two thirds of the world's largest cities are located on them. Estuaries are also important sites for aquaculture, used for the production of salmon, prawns, and shellfish such as mussels and oysters.

Around the world, estuaries are under threat from pollution and overfishing. Pollution is a particular problem as it tends to accumulate in estuaries, where agricultural, urban, and industrial run-off flows in from rivers and is often joined by pollution from ships and sewage. Flood control and water diversion are also problems. The influx of fertilizers and sewage can lead to eutrophication: excess nutrients cause algae to bloom and then decay, removing dissolved oxygen from the water and eventually leading to the development of so-called dead zones.

In some areas, rising sea levels are also causing seawater to intrude more deeply into estuaries—even up into the rivers themselves—with potentially negative impacts on the creatures that live there.

// Ice ages

Every so often, the Earth enters an ice age, when global temperatures drop and glaciers and ice sheets spread out across vast areas of land, reshaping the planet's surface.

Ice ages can last for millions or even tens of millions of years. There have been at least five major ice ages in the Earth's history. The first took place more than two billion years ago; the most recent, known as the Quaternary Ice Age, began about 2.6 million years ago and continues to the present day.

Ice ages alternate with warmer periods during which all of the world's glaciers and ice sheets melt, and the planet is ice free even at high latitudes. These are known as greenhouse periods.

During an ice age, periods known as glacials, when temperatures drop and glaciers and ice sheets advance, alternate with periods known as interglacials, when temperatures warm up and glaciers retreat. The most recent glacial, often referred to informally as the Ice Age, reached a peak about 20,000 years ago, when the global average temperature was about 9°F (5°C) colder than today and in some areas was as much as 40°F (22°C) colder. About 11,000 years ago, a warm interglacial, known as the Holocene, began.

Glacials and interglacials tend to take place in fairly regular repeated cycles, the timing of which is largely governed by changes in the Earth's orbit. Called Milankovitch cycles, these changes affect the amount of sunlight that reaches different parts of the Earth's surface. During the Quaternary, the glacials have typically lasted about 70,000–100,000 years, while the interglacials have lasted about 10,000–20,000 years.

Ice ages typically develop slowly but end more abruptly, as do the glacials and interglacials within them. They can be triggered by a number of different phenomena, including changes in the circulation patterns of the ocean and atmosphere, and changes to the amount of carbon dioxide in the atmosphere.

The restless movement of the world's continents and consequent changes to ocean circulation is a potentially significant trigger, particularly when the flow of warm water from the equator to the poles is blocked. The current ice age was probably triggered by the closure of the gap between North and South America, when the formation of the Isthmus of Panama stopped the exchange of tropical water between the Atlantic and Pacific oceans, significantly altering ocean currents and the circulation of heat around the oceans.

The composition of the Earth's atmosphere can also trigger or end an ice age. For example, the Huronian, the earliest well-established ice age, which took place about 2.4–2.1 billion years ago, is believed to have been caused when atmospheric methane, a powerful greenhouse gas, was eliminated during what is known as the Great Oxygenation Event. Similarly, the evolution of land plants at the beginning of the Devonian period led to a long-term reduction in carbon dioxide levels, which triggered an ice age in the late Palaeozoic (about 340 million years ago). Volcanic eruptions often cause the release of large amounts of carbon dioxide and can thus initiate either interglacial or greenhouse periods.

During glacial periods, continental and polar ice sheets and alpine glaciers in both hemispheres expand significantly in both area and volume. During the Cryogenian period, from 720 to 630 million years ago, glacial ice sheets are thought to have reached the equator, creating a "snowball Earth." At these times, sea levels drop as large volumes of water are frozen in the ice caps. At the height of the last glacial period, the ice sheets were as much as 2.5 miles (4 kilometers) thick and sea levels were about 390 feet (120 meters) lower than they are today. This led to the exposure of the continental shelves and the creation of land bridges that allowed animals to cross areas that would normally be underwater. All told, there was almost 20 percent more land than there is today.

Once the growth of ice sheets is set in motion, positive feedback loops can kick in, causing runaway growth and the beginning of a true ice age. For example, snow and ice increase the Earth's albedo (reflectivity), reducing absorption of the Sun's energy as it is reflected back into space. This causes the temperature to drop, which causes areas of snow and ice to grow.

Right *An artist's impression of the Earth at the last glacial maximum, when vast ice sheets reached deep into Europe.*

// Glaciers and ice caps

Remnants from the last ice age, glaciers and ice caps are slow-moving rivers of ice that have the power to shape landscapes.

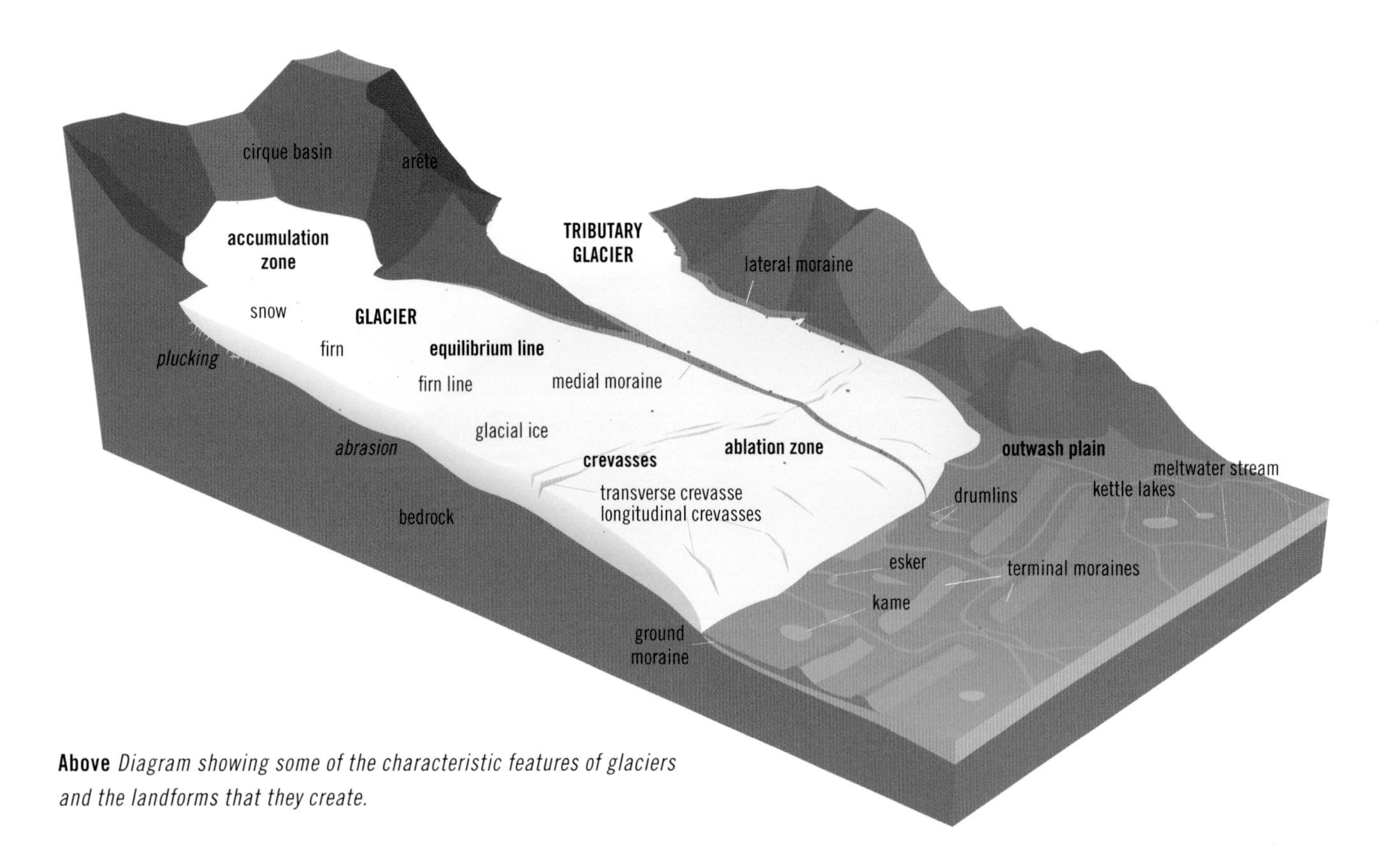

Above *Diagram showing some of the characteristic features of glaciers and the landforms that they create.*

Glaciers are large perennial masses of ice that move slowly over land (often along a mountain valley), driven by their own weight. Dome-shaped glacial masses that flow in all directions and are not constrained to mountain areas are known as ice caps. An area with interconnected ice caps and glaciers is termed an ice field. When the summit of a mountain rises above the ice, it's called a nunatak. Glaciers can be as small as a football pitch or several hundred miles long. Larger masses of ice that cover an area greater than about 20,000 square miles (50,000 square kilometers) are known as ice sheets (see page 44).

Ice caps and glaciers form when fallen snow accumulates to the point where it compresses to become ice. Newly fallen snowflakes evaporate, condense, melt and freeze together to form grains similar in size and shape to sugar crystals. If this snow survives a melting season, it's called firn or névé.

As more snow falls on top, layers of firn build up and their weight further compresses the ice beneath them, squeezing out any air pockets and increasing the ice's density. Eventually, the ice grows so thick that its weight causes the firn grains to fuse into a solid ice mass. The glacier then begins to move under its own weight.

The speed and depth at which fallen snow becomes glacier ice depends on the surrounding temperature. In warm, wet environments, it may take three or four years and less than 33 feet (10 meters) of burial; in much colder climates, on the other hand, it may take several thousand years and depths of about 490 feet (150 meters). As counter-intuitive as this may seem, it occurs because in the warmer, wetter environment, more snow will fall and then melt, so the individual snowflakes turn to ice relatively quickly. In a colder environment, the snow is less likely to ever melt, so the transformation into ice mainly occurs due to compression, which requires the weight provided by a great depth of snow.

Put simply, glaciers flow under the influence of gravity, their sheer weight causing them to move downhill, but the actual process of glacial sliding is poorly understood. Generally, glaciers move very slowly—about 10 inches (25 centimeters) per day—although sometimes they will "surge," moving forward by several feet per day for weeks or months before returning to their previous slow creep.

The ice at the surface of a glacier moves more rapidly than that on its underside, due to friction between the base and the ground. Sometimes there will be a thin layer of water that separates the glacier's underside from the ground, which reduces the amount of friction and thus allows the glacier to flow more rapidly. Subglacial meltwater can form braided streams that discharge at the front of the glacier. Pressure from the overlying ice can sometimes force such streams to flow uphill.

Glaciers that terminate in the ocean are known as tidewater glaciers. These are the source of icebergs. They tend to flow relatively quickly – at a rate similar to that of a surging mountain glacier.

Snowfall (accumulation) feeds glaciers, while melting and the calving off of icebergs (ablation) diminish them. The difference between these processes is known as the glacier's mass balance – if it is positive, with accumulation greater than ablation, the glacier will grow (advance); if it is negative, it will shrink (retreat). The boundary between the accumulation and ablation zones is known as the equilibrium line.

Glaciers currently occupy about 10 per cent of the world's total land area. They occur all over the world at almost all latitudes. Most are located in the polar regions, but there are high-altitude glaciers at or near the equator. Glaciers and ice caps hold about 70 per cent of the world's fresh water. Estimates suggest that if all of the currently extant glaciers were to melt, sea level would rise by about 70 metres (230 feet).

Svínafellsjökull is one of about 30 outlet glaciers that flow out from Iceland's Vatnajökull ice cap, the largest ice cap in Europe by volume.

// Ice sheets

Also known as continental glaciers, ice sheets are masses of glacial ice that cover an area greater than 19,300 square miles (50,000 square kilometers).

Today, the only existing ice sheets are in Antarctica and Greenland, but during the last glacial period much of North America was covered by the Laurentide ice sheet, the Weichselian ice sheet covered northern Europe, and southern South America was covered by the Patagonian ice sheet. Together, today's two remaining ice masses contain about 99 percent of the world's glacier ice; 91 percent is in Antarctica alone.

Covering an area of nearly 5.4 million square miles (14 million square kilometers), the Antarctic ice sheet is by far the largest single mass of ice on Earth, containing more than 6 million cubic miles (25 million cubic kilometers) of ice with a mean thickness of about 6,000 feet (1,829 meters)—enough to raise global sea levels by 190 feet (58 meters) if it melted completely. It is divided in two by the Transantarctic Mountains. The larger East Antarctic ice sheet, which covers about 3.9 million square miles (10.2 million square kilometers) and holds the thickest ice on Earth, at 15,700 feet (4,800 meters) deep, rests on a major landmass. Parts of the West Antarctic ice sheet, on the other hand, are more than 8,200 feet (2,500 meters) below sea level. It features two large ice shelves, each more than 193,000 square miles (500,000 square kilometers) in size.

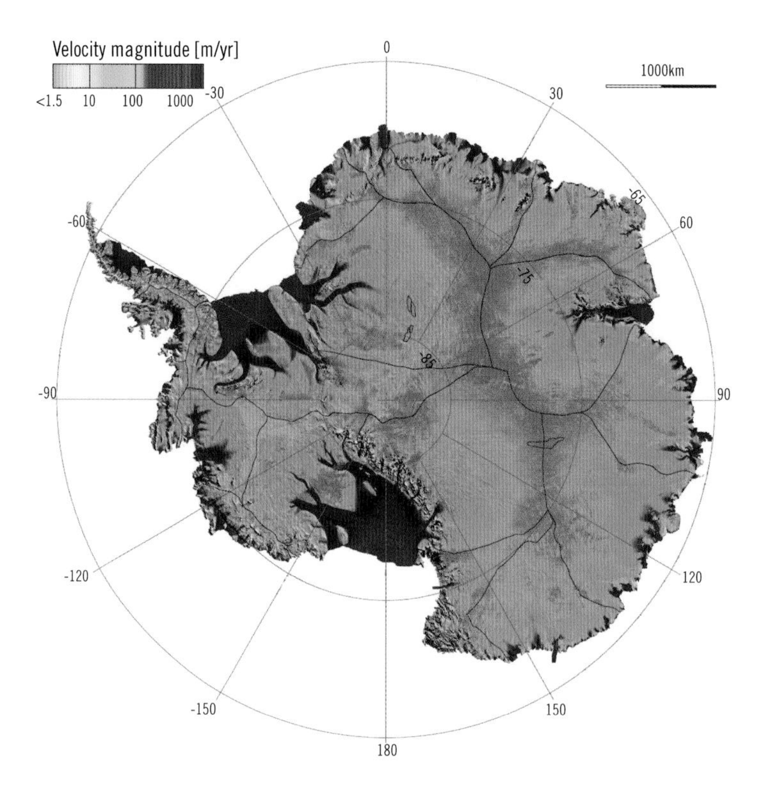

Above *Map of the speed and direction of ice flow in Antarctica. The black lines show the major divisions within the ice sheets, as well as the subglacial lakes in the continent's interior. The ice flows slowly in central areas and speeds up as it reaches the coast as it is funneled into outlet glaciers and out onto the floating ice shelves.*

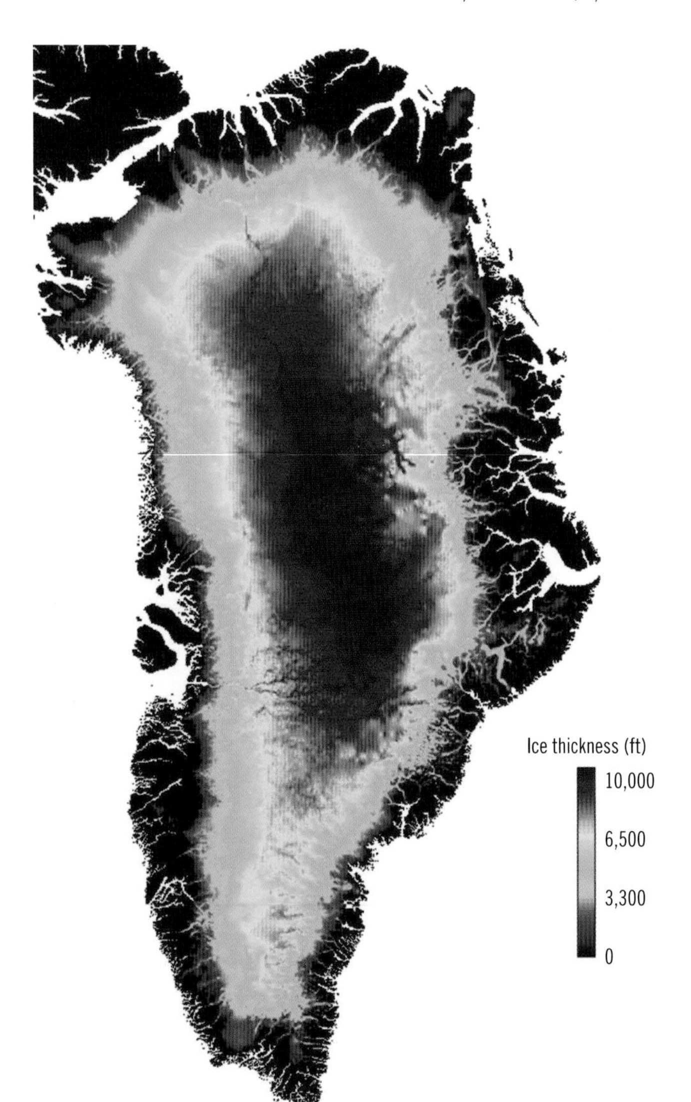

Left *Ice thickness, Greenland. The Greenland Ice Sheet is the world's second largest body of ice, covering an area of 656,374 square miles (1.7 million square kilometers) or about 80 percent of the surface of Greenland. Such is the weight of the ice, which is more than 1.9 miles (3 kilometers) deep at its thickest point, that the bedrock at the center of Greenland has been depressed almost to sea level.*

Above *The summits of mountains, known as nunataks, protrude from the surface of the Greenland ice sheet.*

The Greenland ice sheet covers an area of about 656,000 square miles (1.7 million square kilometers), or roughly 82 percent of the island of Greenland, and has a total volume of almost 720,000 cubic miles (3 million cubic kilometers). It resembles an elongated dome or ridge with two summits, one of which is almost 11,000 feet (3,300 meters) high. If the Greenland ice sheet were to melt completely, sea levels would rise by 24 feet (7.2 meters).

Over most of Greenland's interior, the bedrock's surface is near sea level, but a ring of mountains runs around the island's periphery, largely confining the ice sheet and restricting it from forming large ice shelves. Instead, large tongue-like protrusions of ice known as outlet glaciers flow out through valleys between the mountains before calving off icebergs into the ocean. At its terminus, one of these, the Jakobshavn Glacier, flows at a rate of up to 72 feet (22 meters) per day.

Like glaciers, ice sheets are in constant motion, slowly flowing downhill under their own weight. In the interior, the ice moves very slowly—a few inches or feet per year—because the surface slope is small and the ice is very cold. The rate of flow increases towards the ice sheet's margins, eventually reaching rates of up to half a mile (0.8 kilometers) per year.

In general, the flow of ice in an ice sheet is directed from the center outward. However, at the sheet's edges, where the ice is thinner, the underlying topography may come into play, creating outlet glaciers.

Geothermal heat and friction between the moving ice and the bedrock warm the ice sheet's base, causing some areas to melt. The melt water lubricates the ice sheet, allowing it to flow more easily and producing rapidly flowing channels known as ice streams.

As with valley glaciers, ice sheets depend on the accumulation of new snow to offset the loss of ice at their edges and maintain a stable mass. They mostly lose mass through surface melting, evaporation, wind erosion (deflation), iceberg calving, and melting at the base of ice shelves by seawater. Global warming has greatly increased the rate at which the Greenland ice sheet is melting; estimates suggest that it is currently losing about 200 gigatons (180 gigatonnes) of ice per year.

The situation in Antarctica is more complex. In some regions—mostly in the interior—ice thickness has increased as a result of heavy snowfall. However, this growth is dwarfed by the ice loss around the margins, where ocean warming is causing the ice shelves to melt from below. Since the mid-1990s, Antarctic ice shelves have lost nearly 4.4 trillion tons (4 trillion tonnes) of ice. Although this does not contribute to sea-level rise directly, as they are already floating in the ocean and displacing water, the ice shelves slow down the flow of ice off the continent and into the ocean, so their loss may accelerate other ice loss. Overall, the Antarctic ice sheet is melting at a rate of about 118 gigatons (107 gigatonnes) of ice per year.

// Glacial landforms

The awesome, implacable power of moving ice has shaped landscapes all over the Earth.

Glacial landforms tend to fall into two types: those created by deposition and those created by erosion.

At the base of most temperate glaciers is a layer that contains rock debris. It may be several inches to a few feet thick. The gathering of material from the bedrock by a glacier is known as glacial plucking or quarrying. This material, which can range in size from silt to boulders, is dragged across the underlying bedrock by the glacier's movement, potentially eroding away tens of feet of rock and sculpting the landscape into characteristic forms.

Smaller erosion-related features vary according to the size of the material being dragged over the bedrock. Fine material known as rock flour can polish stone, while larger rocks can create long, deep gouges known as striations.

In the upper reaches of a valley glacier, erosion is generally greater than deposition, but closer to the terminus or snout—the downslope end of the glacier—deposition outweighs erosion. Hence the high areas of glaciated mountain ranges are dominated by erosional landforms. Alpine landscape features such as sharp mountain peaks, steep-sided valleys, lakes and waterfalls are mostly the product of previous periods of glaciation.

At the top of most glacial valleys is a cirque, a semi-circular amphitheater-shaped hollow facing down the valley. The cliff at the top of a cirque is called the headwall. A basin often forms at the base, perhaps containing a lake termed a tarn. Between two cirques or neighboring parallel valleys, there may be jagged, knife-edge ridges known as arêtes; a low saddle on an arête is called a col. Where several cirques abut, they may leave a steep, pointed peak known as a horn.

Below *As a valley glacier moves downhill, friction causes the ice at its base to melt. This water seeps into cracks in the bedrock. When it refreezes, it expands, causing pieces of rock to break off and become trapped in the glacier, a process known as plucking. As the rock pieces are dragged over the bedrock, they gouge out striations and wear away protrusions, a process known as abrasion.*

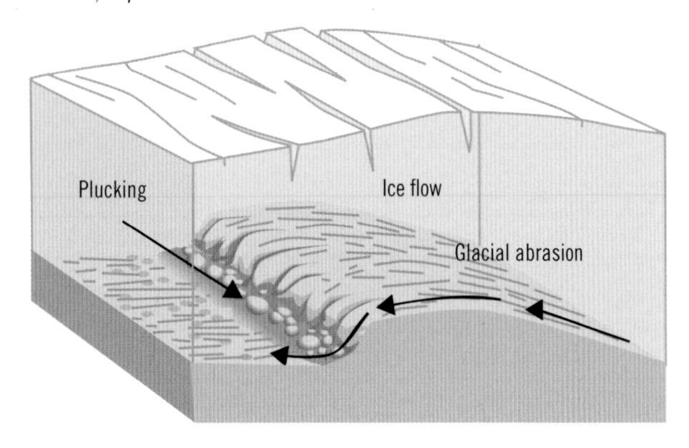

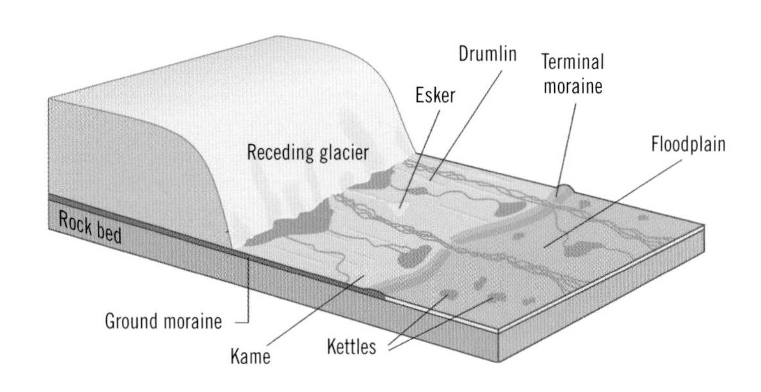

Above *When glaciers recede, they leave behind a series of characteristic depositional landforms that may include moraines, kames, drumlins, eskers and kettles.*

Because they are in contact with large areas of the valley floor and are so powerful, glaciers tend to erode a much larger surface area than streams and rivers, which only wear away a narrow line along the valley's lowest part. Thus streams and rivers tend to form V-shaped valleys while glacial valleys are U-shaped, with relatively wide and flat bottoms and steep walls. Knobs or hills of bedrock left behind while the surrounding material is abraded away are known as roches moutonnées or rock drumlins.

Large mountain glaciers are often fed by smaller tributary glaciers. The former tend to gouge out deeper valleys and, when the ice retreats, the bases of the tributary valleys are left high up the wall of the main valley. Known as hanging valleys, these features often support waterfalls. Where a glacially formed coastal valley has been flooded by the rising sea it's known as a fjord.

Ice forming in cracks in cliffs causes chunks of rock to break off, a process known as frost shattering. Piles of rock formed at the base of cliffs by this process are known as talus. Sometimes, if enough ice freezes around an area of talus it will begin to move downhill, becoming what's known as a rock glacier.

When a stream flows down the upper surface of a glacier, it can create a channel bounded at the sides by ice. If the glacier retreats, sediment deposited by the stream is left behind as a sinuous steep-sided ridge 65–100 feet (20–30 meters) high and possibly hundreds of miles long, known as an esker.

In a glacier's lower reaches, temperatures are typically higher than at the top and significant melting may occur. The water flows out in channels beneath, within or beside the glacier and often carries sediment and other debris.

At the terminus, the material carried by the glacier is deposited as the ice around it melts, forming a rocky and/or muddy blanket over the terminus. Glacial debris deposited directly by the ice is known as till; deposits of sediment from meltwater streams, which tend to accumulate in lakes below the glacier's terminus, are called glacial outwash or glacial drift. Till may often accumulate in moraines—ridges several hundred feet high and hundreds of yards wide. Boulders of a different rock type to the bedrock on which they are deposited are called erratics.

Till can also be arranged in drumlins, small elongated hills aligned in the direction of glacier flow. They can occur in large numbers in so-called drumlin fields. And it may also be deposited in irregularly shaped mound-like hills known as kames.

Below *Glacial moraines near Lake Louise in Alberta, Canada.*

// Lakes

Found in their millions all over the world, lakes feature on every continent and in every kind of environment, from desert to tundra.

Above *The lake in the active Kawah Ijen stratovolcano in East Java, Indonesia, the world's largest highly acidic crater lake. More than half a mile (0.8 kilometers) wide, the lake's turquoise color is the result of metals dissolved in the acidic water, which has a pH as low as 0.5.*

A lake is a body of water that is surrounded by land, apart from where a river or stream feeds or drains it. Most of the world's lakes are freshwater and are found at higher latitudes in the Northern Hemisphere—the US state of Alaska alone boasts more than three million lakes with a surface area greater than 20 acres (8 hectares). Small lakes are often referred to as ponds, but there is no accepted definition for the term.

Most lakes are fed by rivers and/or streams and typically also have at least one natural outflow, allowing for drainage of excess water, so the water level remains roughly the same over time. All freshwater lakes are of this "open" type. "Closed" lakes, which do not have a natural outflow and only lose water by evaporation, underground seepage, or both, are known as endorheic lakes and are usually saline (see page 50).

Lakes are typically characterized according to origin. There are 11 major types: tectonic lakes, volcanic lakes (occupying depressions created by volcanism), landslide lakes (formed where debris from a landslide blocks the flow of a stream or river; such lakes tend to be relatively short-lived), glacial lakes, solution lakes (formed when underlying bedrock dissolves, often in karst areas), fluvial lakes (created by running water), aeolian lakes (produced by wind action, generally in arid environments), shoreline lakes (usually created by the blockage of an estuary or the formation of a sand spit), organic lakes (created by plants, or animals such as beavers), anthropogenic (man-made) lakes, and meteorite (extra-terrestrial-impact) lakes. Oxbow lakes are a type of fluvial lake that forms when the deposition of sediment causes a crescent-shaped river meander to be cut off from the main flow (see page 30).

The bulk of the world's lakes, especially those in the Northern Hemisphere, were formed during the last glacial

Below *Lakes are typically classified according to how they formed.*

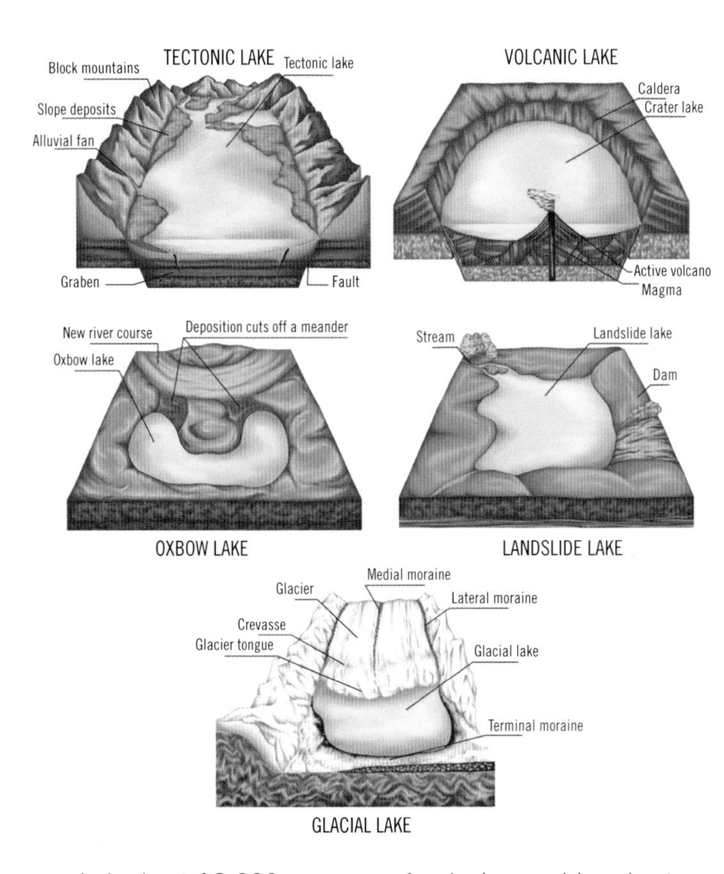

period, about 18,000 years ago. As glaciers and ice sheets moved across the landscape, they gouged out basins of various sizes; when they retreated, rain and meltwater filled the basins. Glacial moraine deposits also dammed rivers to form lakes. In some areas, glaciation led to the formation of tens of thousands of lakes that cover as much as three quarters of the total land area. The Great Lakes in North America were created by glacial action.

At almost 1.2 miles (2 kilometers) deep, Russia's Lake Baikal is the world's deepest lake. It is also the world's oldest (more than 25 million years) and the largest freshwater lake by volume (5,700 cubic miles/23,600 cubic kilometers). It is an example of a tectonic lake, as is the Caspian Sea—the world's largest lake, with an area of more than 143,000 square miles (370,000 square kilometers)—which was created when its basin was cut off from the ocean by tectonic uplift.

All lakes have finite lifespans. Over hundreds or thousands of years, they fill with sediment and organic material and begin to shrink, slowly turning into marshes or swamps (see page 78), at which point the drying process slows down. Eventually, however, they will turn into dry land.

Intermittent, ephemeral or seasonal lakes disappear seasonally. Often found in karst areas, they typically only fill following above-average precipitation.

Larger lakes tend to be thermally stratified. One of the reasons for this layering is the unusual relationship between water's temperature and density: at sea level, fresh water is densest at about 39°F/4°C (this is why frozen lakes usually have liquid water at the bottom, allowing fish to survive the winter). The cool, dense bottom layer of water is called the hypolimnion. At the surface, the Sun heats the water, forming a less dense top layer known as the epilimnion. The central layer (the metalimnion) forms a thermocline between the top and bottom layers. The changing of the seasons can cause these layers to mix, a process known as turnover. In winter, the epilimnion may be colder than the hypolimnion and may even freeze.

Lake Baikal in Russia is the world's deepest and oldest lake, and the largest freshwater lake by volume. It is located on a plate boundary and formed more than 20 million years ago when a rift valley opened up as the plates diverged.

THE INCREDIBLE DISAPPEARING LAKE

On June 3, 2005, Lake Byeloye in Russia's Nizhny Novgorod Oblast vanished in just a few minutes. Local government officials suggested that the sudden disappearance was due to a shift in the soil beneath the lake that allowed the water to drain into the adjacent Oka River. Melting permafrost is thought to be causing similar disappearances of hundreds of large Arctic lakes across western Siberia.

SALINE LAKES AND SALT PANS

When lakes lack an outlet for their water, salt builds up and they can eventually become saltier than the ocean. Defined as land-locked water bodies that have salinities in excess of 3 grams per liter, saline or salt lakes are found on all of the world's continents, including Antarctica, typically in arid and semi-arid regions. Those that have a higher concentration of salt than seawater are known as hypersaline lakes.

Saline lakes with a high concentration of carbonates and bicarbonates are called soda lakes. They are found in volcanic craters in Africa's Eastern Rift Valley and on the high Altiplano of South America. In both of these regions, the soda lakes are inhabited by large flocks of flamingos.

Generally, saline lakes form in endorheic basins—that is, basins that do not have an outlet. Water flows in from rivers and streams or snowmelt but is only lost through evaporation. The dissolved salts that flow into the lake are left behind when the water evaporates, building up over time and making the water increasingly salty. In some cases, saline lakes form when water bodies that were once connected to the ocean become land-locked.

Human activity can also lead to the salinization of a lake. The diversion of rivers for irrigation caused the Aral Sea in Central Asia to shrink significantly and its salt levels to rise, leading to a dramatic drop in biodiversity in the lake and the collapse of the local fishing industry.

About 45 percent of all inland water is saline. The world's largest lake, the Caspian Sea, and its lowest lake, the Dead Sea, are both saline lakes. Many of the world's highest lakes, including those on the Altiplano of South America, are also saline. Don Juan Pond in Antarctica's McMurdo Dry Valleys has an estimated concentration of calcium carbonate of more than 600 parts per thousand. The salt concentration is so high that the pond's freezing point is about -62°F (-52°C).

The high salt levels in and around saline lakes limit the types of organisms that can live in or even close to them, so they typically have low biodiversity.

Below *Tanzania's Lake Natron is a soda lake filled with a caustic, highly alkaline brine. The high levels of natron (sodium carbonate decahydrate) and trona (sodium sesquicarbonate dihydrate) in the lake are the result of the weathering of the surrounding bedrock, which is composed of alkaline lavas that contain significant amounts of carbonate, and high levels of evaporation.*

The Dead Sea, which is located in the Jordan Rift Valley at the world's lowest elevation on land. It is one of the world's saltiest water bodies, with an average salinity of about 300 parts per 1,000, or 30 per cent.

Hexagonal salt deposits on the surface of Bolivia's Salar de Uyuni, the world's largest salt flat. The salt crust is several metres deep and covers a layer of brine that is exceptionally rich in lithium.

SALT PANS

When the rate of evaporation of a saline lake outstrips the rate of water inflow, the lake will shrink and eventually dry up completely to form a salt pan, also known as a salt flat or playa. The layer of salt crystals and other minerals on the salt pan's surface will often turn it white. Many salt pans act as ephemeral lakes, filling with water periodically, usually following a significant storm somewhere in the lake's drainage basin.

The world's largest salt pan is the Salar de Uyuni in the Bolivian Andes, which covers an area of 10,852 square kilometres (4,190 square miles). Its salt deposits contain valuable minerals, including sodium, potassium, lithium and magnesium, which support a lucrative mining industry.

The amount of salt present in a salt flat can be substantial; the Bonneville Salt Flats in Utah are 1.5 metres (5 feet) thick at the centre and contain about 133 million tonnes (147 million tons) of salt.

// Inland seas

When sea levels are high, salt water may flow into the interiors of continents, creating shallow but extensive water bodies known as inland, epeiric or epicontinental seas. They remain connected to the ocean or adjoining seas by one or more straits.

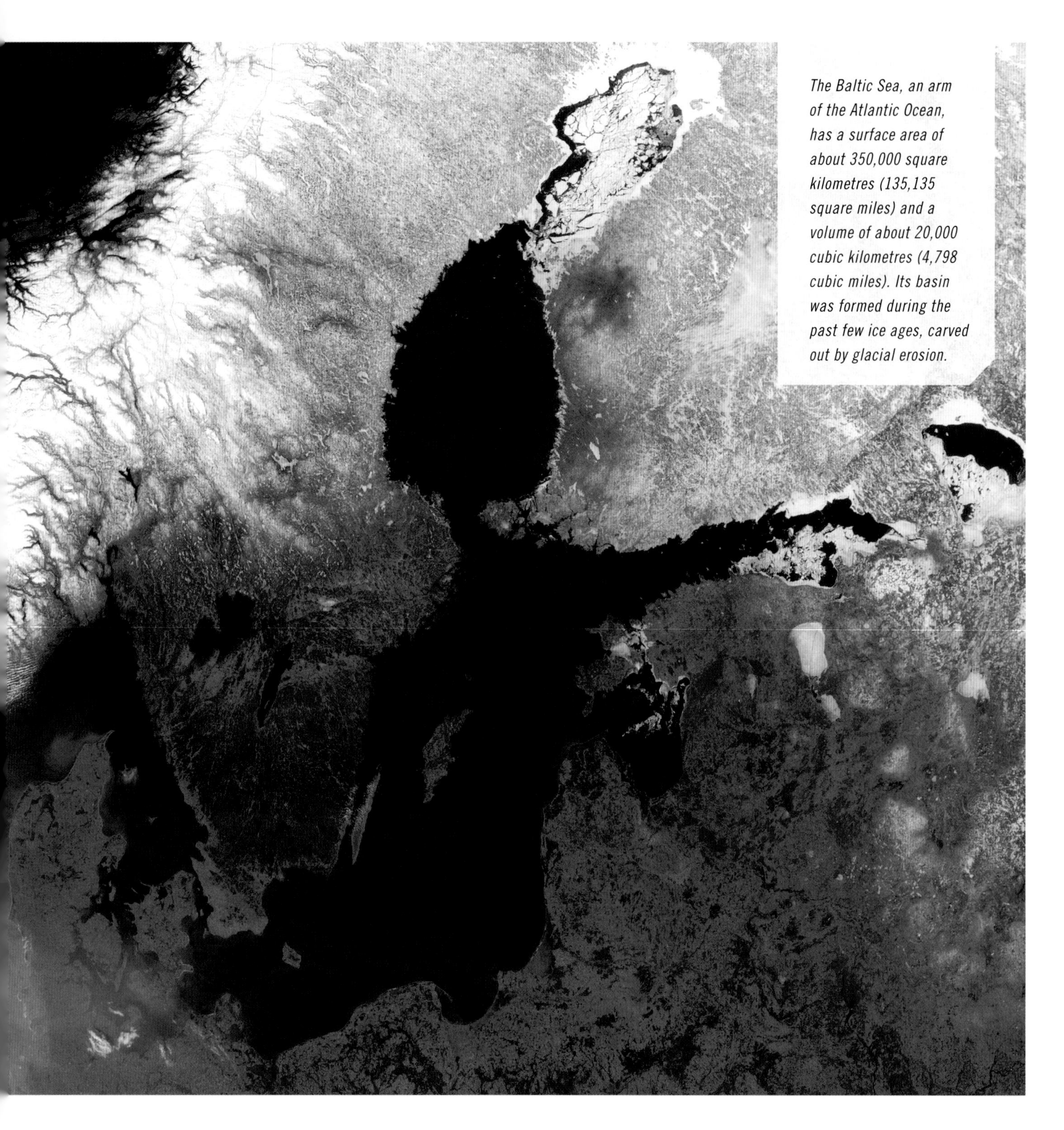

The Baltic Sea, an arm of the Atlantic Ocean, has a surface area of about 350,000 square kilometres (135,135 square miles) and a volume of about 20,000 cubic kilometres (4,798 cubic miles). Its basin was formed during the past few ice ages, carved out by glacial erosion.

Today, the continents are relatively high and sea levels are relatively low, so there are few true inland seas. Current examples include the Baltic Sea in Europe, Hudson Bay in North America, and the Marmara Sea in Turkey (there is some debate as to whether the South China Sea and the Persian Gulf also qualify as inland seas).

Inland seas were much more common during previous eras, when sea levels were often much higher than they are now. They were where the bulk of the Earth's exposed sedimentary rocks were laid down and where most of the fossil record and the world's petroleum were created.

During the Cretaceous period, the Eromanga Sea covered much of eastern Australia and the Western Interior Seaway extended from the Gulf of Mexico deep into present-day Canada. More recently, during the Oligocene and Early Miocene epochs, the sea flooded much of Patagonia in South America, at one point possibly linking the Pacific and Atlantic oceans. And during the Cretaceous period, the low plains of modern-day northern France and northern Germany were also flooded by an inland sea, within which the chalk that gave the geological period its name was deposited.

Inland seas tend to be greatly influenced by the adjacent land. In the Baltic, for example, the surface salt water is strongly diluted by water flowing in from the nine rivers that empty into the sea. Their influence makes the Baltic the world's largest brackish sea. In other cases, the water may be highly saline due to a drier inland climate and consequently higher rates of evaporation.

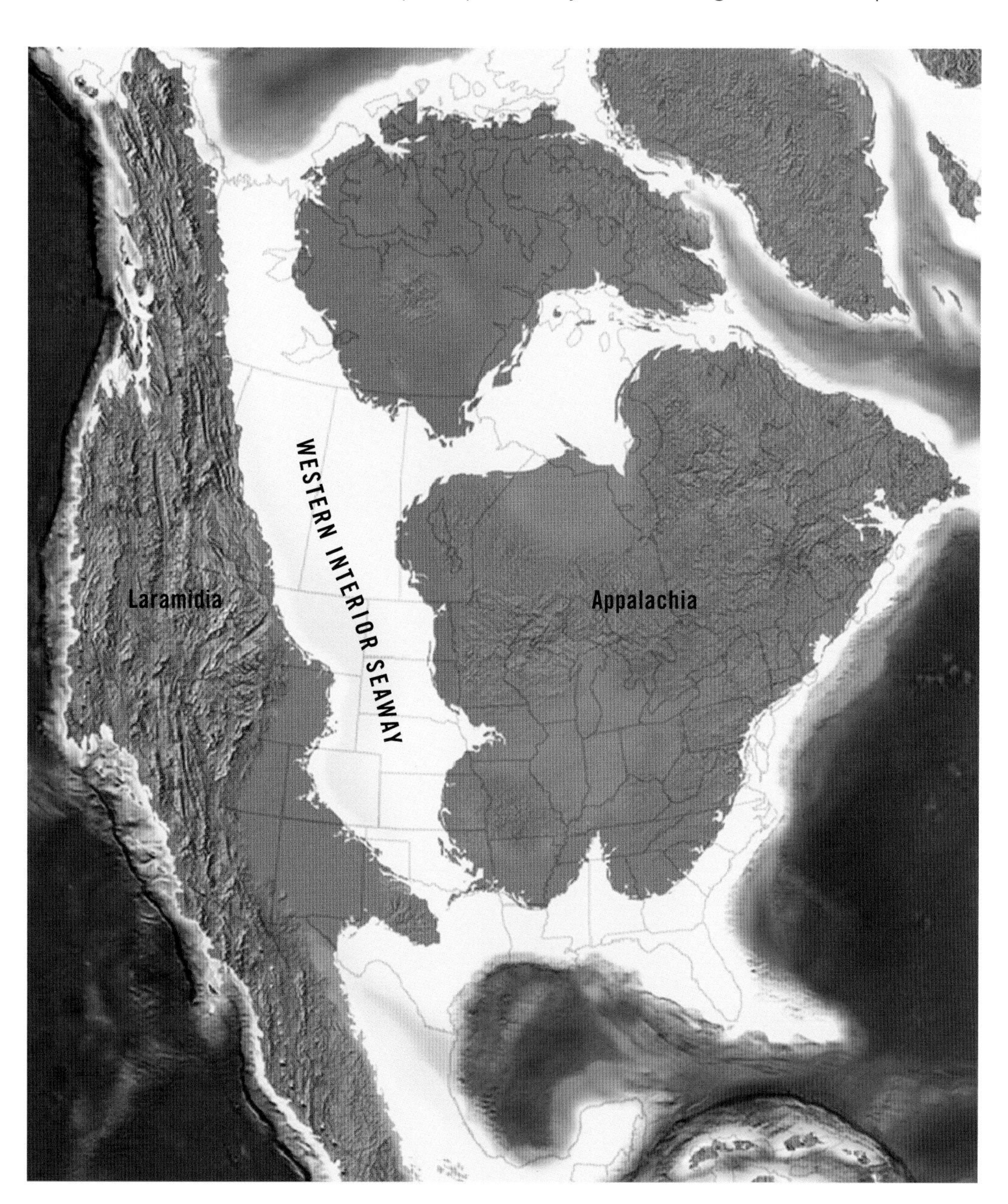

THE INLAND SEA THAT ISN'T

Despite being the world's largest inland water body and having a salinity about a third that of seawater, the Caspian Sea in Central Asia is not technically an inland sea because it lacks a connection to an ocean. Like the Black Sea, the Caspian Sea is a remnant of the ancient Paratethys Sea. It became landlocked about 5.5 million years ago due to a combination of tectonic uplift and falling sea levels.

Left *The Western Interior Seaway was an inland sea that divided the continent of North America into two landmasses. As much as 2,600 feet (792 metres) deep, 600 miles (966 kilometers) wide and 2,000 miles (3,200 kilometers) long, it existed during the mid- to late Cretaceous period and into the early Paleogene.*

// Groundwater and aquifers

Hidden beneath the Earth's surface is a vast storehouse of water held in enormous subterranean aquifers (water-bearing rocks).

In areas where the ground is permeable, water soaks down until the soil and rock become saturated. The upper surface of this saturated zone is known as the water table. The level of the water table moves up and down naturally, depending on the amount of rainfall.

The water held in the saturated soil and rock beneath the water table is known as groundwater. Groundwater accounts for about 98 percent of global fresh water in liquid form and provides at least half of the world's drinking water and more than 40 percent of water for irrigation.

If the material in which the groundwater is held readily transmits that water to wells and springs, it is known as an aquifer. Aquifers can form in numerous different types of sediment and rock, including gravel, sandstone, conglomerates and fractured limestone.

Although water will move freely within many aquifers, groundwater is not like an underground river or lake: the water is held in the tiny pores and cracks within rocks and sediments—much like water in a sponge. The speed at which it moves within the aquifer depends on the aquifer's permeability. In some aquifers, groundwater moves several feet in a day; elsewhere it may only move a few inches in a century.

Simple wells work by digging down below the water table and into an aquifer. Water then seeps out of the surrounding rocks and sediments, and into the well, like water filling a hole dug in beach sand. The level of the surface of the water in the well will be the same as that of the water table. If too much

THE GREAT ARTESIAN BASIN

Up to 9,800 feet (3,000 meters) deep and stretching over 656,000 square miles (1.7 million square kilometers), the Great Artesian Basin underlies about a fifth of Australia. It's the largest and deepest artesian basin in the world and is estimated to contain 15,600 cubic miles (64,900 cubic kilometers) of groundwater.

Below *Map of global groundwater resources.*

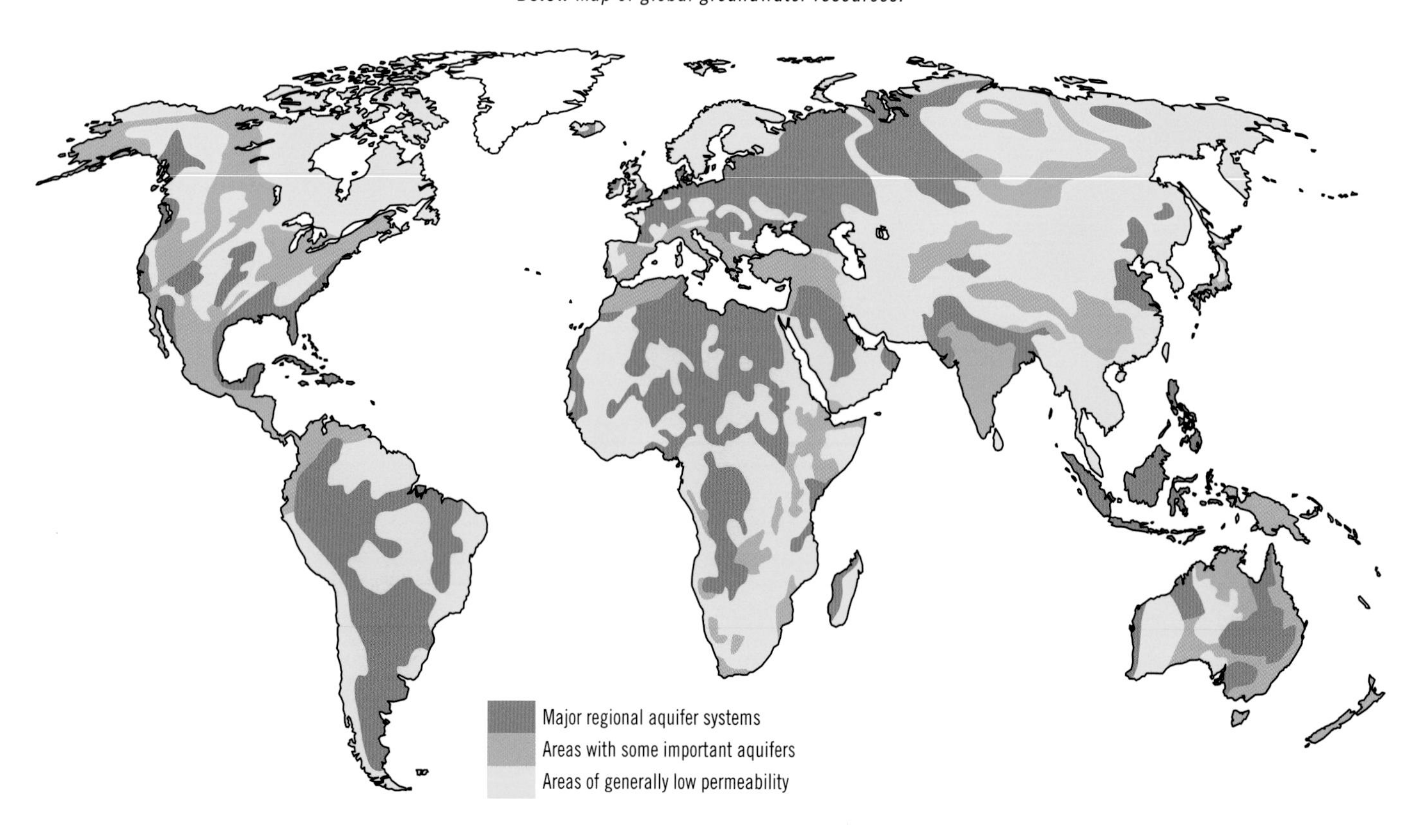

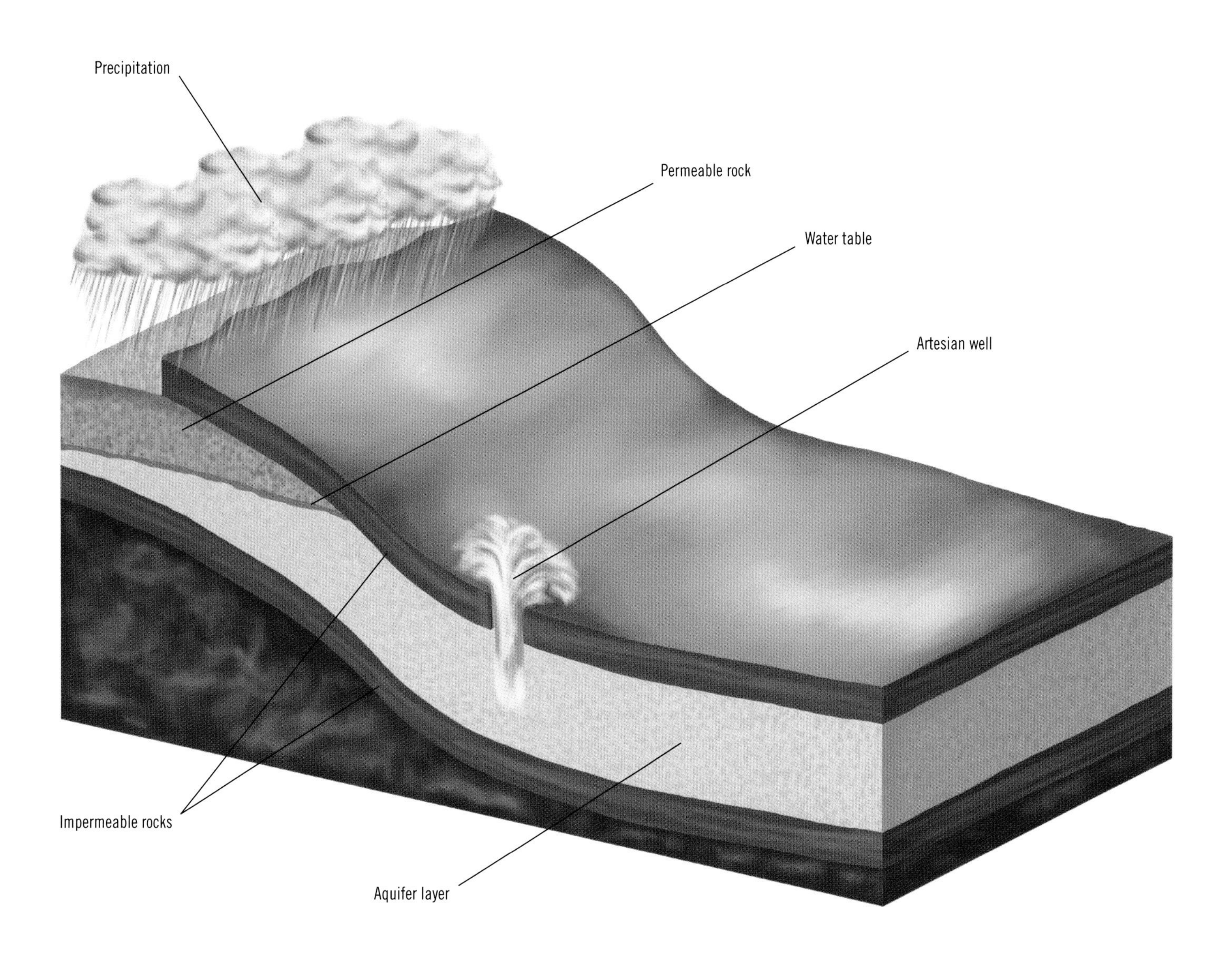

Above *Sometimes, the porous rocks that make up an aquifer are sandwiched between layers of less porous rock; this is known as a confined aquifer. If these rock layers are tilted, water will move diagonally down through the aquifer, often emerging in low-lying ground as a spring or a soak. The water in a confined aquifer can also become pressurized. If a well is dug into this type of aquifer, the pressure will push the water up to the surface without the need for a pump. This is known as an artesian well.*

groundwater is drawn up to the surface, it can cause the water table to drop below the level of the well, causing it to "go dry."

When water is drawn from an aquifer faster than it is replenished by rainfall, a process known as recharging, the groundwater can become depleted. This can lead to subsidence of the overlying land. In Florida, over-extraction of water for strawberry farming has resulted in the opening up of hundreds of sinkholes, while in California's San Joaquin Valley, the ground level has dropped more than 26 feet (8 meters) since the 1920s.

FRESHWATER LENSES

On some small coral or limestone islands and atolls, there is a convex layer of fresh groundwater that floats on top of the denser seawater. This freshwater aquifer is known as a lens. Freshwater lenses are vital sources of drinking and agricultural water for many island communities. Because the atolls that support freshwater lenses are often only a few feet above sea level, they are at significant risk of inundation due to rising sea levels.

// Rocks, minerals, and gems

The Earth's crust is made up of a dizzying array of minerals, which make up the rocks we see around us—and some of which we have attached a high value to.

Minerals are naturally occurring substances formed through geological processes. They have a characteristic chemical and physical composition, and usually have a crystalline structure. There are more than 2,800 naturally occurring minerals, which range in composition from pure elements and simple salts to complex silicates that occur in thousands of different forms.

Rocks are aggregates of minerals. The Earth's crust consists of three types of rock: igneous (those that have solidified from a molten state), sedimentary (those that have formed as layers of sediment have been laid down and compacted), and metamorphic (those that have been transformed from other types of rock by temperature and/or pressure). The process of change among the different rock types is referred to as the rock cycle.

There are three different types of metamorphism: contact metamorphism, where heat from magma intruding into existing rock formations causes the rocks to recrystallize into new minerals; regional metamorphism, which happens over a wider area, often along the edges of colliding tectonic plates; and pressure metamorphism, where the weight of overlying rock transforms deeply buried sediments. Contact metamorphism is usually restricted to relatively shallow depths.

The Earth's crust is made up of roughly 65 percent igneous rock, 27 percent metamorphic rock and 8 percent sedimentary rock. Igneous rocks can be divided into two main types: plutonic or intrusive, which form when magma cools and crystallizes slowly within the Earth's crust (for example granite); and volcanic or extrusive, which form when magma reaches the surface as either lava or tephra before cooling (for example basalt). The most common

Below *Minerals come in a wide range of forms and colors, depending on their chemical composition and structure.*

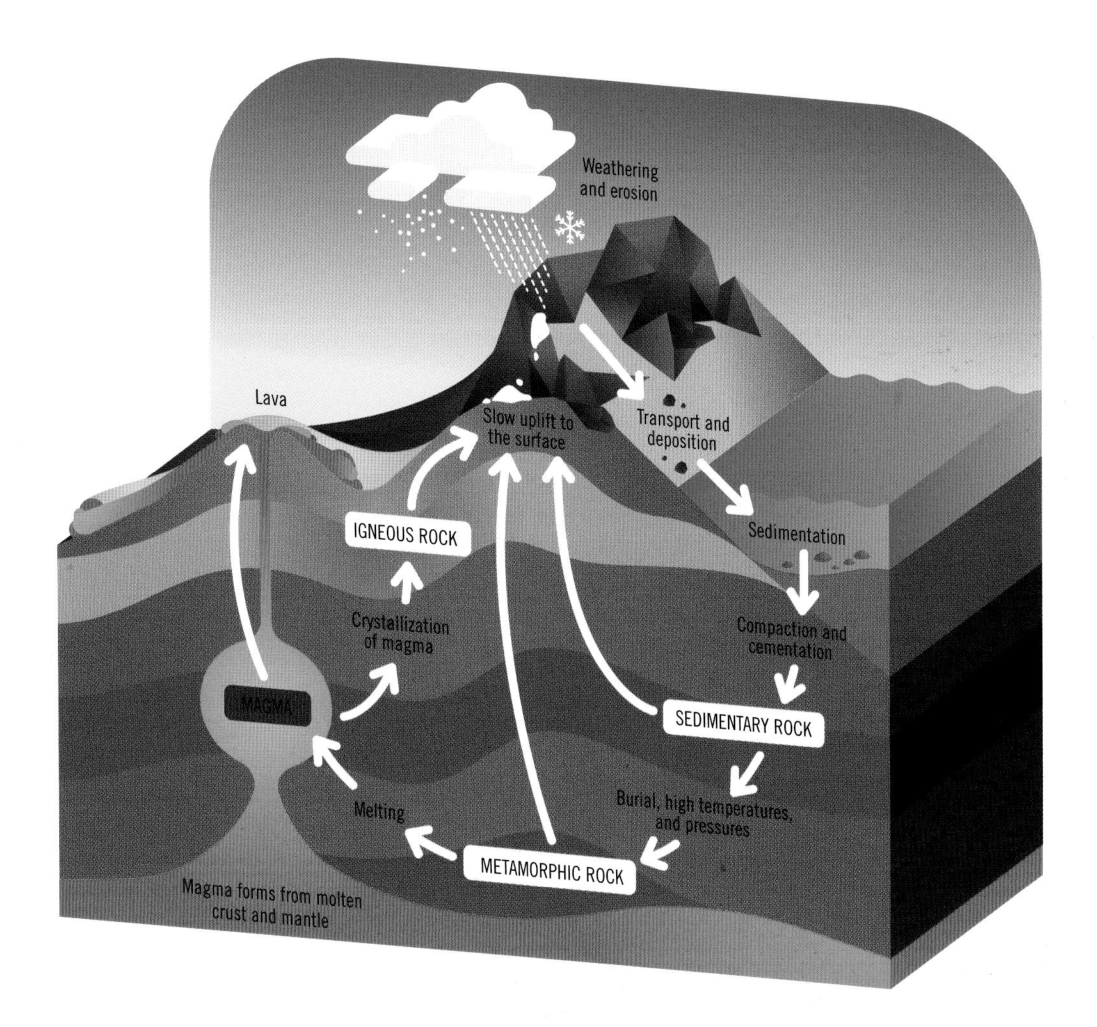

Above *The rock cycle. Over millions of years, a series of processes such as weathering, erosion, sedimentation, compaction, uplift, and melting create and transform the rocks in the Earth's crust.*

sedimentary rocks are shales (82 percent), sandstone (12 percent) and limestone (6 percent). Many sedimentary rocks contain fossils.

Basalt is the most common igneous rock on Earth (it accounts for about 90 percent of all such rock; oceanic crust is 99 percent basalt) and underlies more of the planet's surface than any other rock type. It forms when lava rich in magnesium and iron rapidly cools and solidifies. Continental crust is dominated by granite and similar rocks, known as granitoids.

The most common class of minerals, and the major components of most rocks, making up about 95 percent of the Earth's crust, are silicates, which are compounds that contain silicon and oxygen, and include quartz, mica, and clay minerals. Among the other major mineral classes are oxides, sulfides, sulfates, carbonates, and halides.

Gems are specific minerals (or combinations of minerals and occasionally organic materials) to which humans have attached a value. They are typically cut and polished for use in jewelry and other decorations.

Most gemstones form in the crust; only two—diamond and peridot—form under extremely high pressure in the mantle, about 95 miles (150 kilometers) below the surface. They are then brought to, or near to, the surface during volcanic eruptions (in what are known as kimberlite pipes) or through erosion or mountain building.

Gemstones can be divided into four groups according to their method of formation: igneous, hydrothermal, metamorphic, and sedimentary. Igneous gems form when magma solidifies. If it reaches the surface, magma solidifies as lava, but if it cools more slowly within the crust, it can crystallize and form minerals. Pegmatites, for example, are created when magma is forced into crevices and other cracks and then slowly solidifies. They include topaz, aquamarine, and tourmaline. Hydrothermal gems form when hot water that is super-saturated with minerals is forced into cracks and cavities in the crust, where the solution cools and the minerals crystallize. Examples include emerald and amethyst. Like metamorphic rocks, metamorphic gemstones form when minerals are subjected to high pressures and temperatures, often in subduction zones. Most gemstones are formed in this way. Examples include jade, lapis lazuli, ruby, sapphire, and garnet. Sedimentary gemstones form when surface minerals dissolve in water that then seeps into cavities and cracks, where the minerals recrystallize. Examples include opal, turquoise, malachite, and azurite.

// Metals

Created within stars, metals make up the Earth's core and are also found in scattered deposits in the crust.

Above *Native gold in quartz and granite. Deposits such as this are created when superheated water that contains dissolved gold and silicon dioxide is forced into cracks in the rocks of the crust and the minerals crystallize, forming gold-bearing quartz.*

Metals are materials that conduct electricity and heat relatively well, and are typically malleable (that is, they can be hammered into thin sheets) or ductile (they can be drawn into wires) crystalline solids. About 95 of the 118 elements in the periodic table are metals (the total depends on exactly how a metal is defined) and about a quarter of the Earth's crust is made of metals by weight; roughly 80 percent of these are light metals such as sodium, magnesium, and aluminum.

How metals are created depends on their atomic weight. Those up to and including iron in the periodic table are largely created through a process known as stellar nucleosynthesis, in which lighter elements undergo successive fusion reactions inside stars, forming heavier elements with higher atomic numbers. Heavier metals are largely created by what is known as neutron capture, in which lighter elements are bombarded with neutrons within stars.

When stars reach the end of their lifetimes, they may eject some of their mass, explode as supernovas or collapse to form extremely dense neutron stars that then collide with other neutron stars. Each of these processes provides the source for the metals found in the universe.

Some of this metal-bearing material eventually coalesced to form the Earth. During this period of the Earth's early formation, molten iron sank to the planet's center, taking most of the precious metals with it. Those that are now present in the Earth's crust and mantle are the result of a later bombardment of meteorites, which took place more than 200 million years later and involved about 22 billion billion tons (20 billion billion tons) of asteroidal material.

Most metals are either lithophiles (rock-loving) or

Above *An iron-rich fragment of the Campo del Cielo meteorite, thought to have been part of a meteor with a diameter of 13 feet (4 meters) that fell to Earth about 4,500 years ago.*

chalcophiles (ore-loving). Lithophile metals, which include magnesium, aluminum, and titanium, combine easily with oxygen and are mostly found as relatively low-density silicate minerals. Chalcophile metals, which include lead, copper, and silver, are usually found in insoluble sulfide minerals. Because they are denser than minerals that contain lithophile metals, they sank lower into the crust before it solidified, so they now tend to be less abundant at the surface. There are also a number of siderophile (iron-loving) metals, including gold, platinum, and iridium. These are relatively rare in the crust as they were dragged down into the core during the Earth's formation; deposits in the crust are the result of the meteorite bombardment mentioned above.

Most metals occur in nature as ores—rocks or sediments that contain one or more valuable minerals. They are usually found in compounds with non-metals—typically as oxides, sulfides, or silicates—but a few, such as copper, gold, platinum, and silver, often occur in a relatively pure state because they don't readily react with other elements. These are known as native metals. Ore deposits are always mixed with unwanted or commercially valueless rocks and minerals, known as gangue.

Ore bodies are formed by a number of different geological processes collectively referred to as ore genesis. There are three major types of ore genesis: internal processes, when ores accumulate due to geological activity, such as when they are brought to the surface in volcanic eruptions; hydrothermal processes, in which ores accumulate around hydrothermal vents due to seawater circulating through cracks in the Earth's crust; and surficial processes, when ores accumulate due to processes such as erosion that take place on the Earth's surface. Ores can also arrive in meteorites, which are particularly high in iron. The Earth effectively contains a finite amount of ore because ore genesis is extremely slow—typically taking millions of years.

Hematite iron ore travels on a conveyor belt at the Corumbá mine in Mato Grosso do Sul, Brazil. An iron oxide, hematite is one of the most abundant minerals on the Earth's surface and in the upper layers of the crust.

// Hydrocarbons

Under certain conditions the long-dead remains of plants and animals are transformed into energy-rich substances known as hydrocarbons—the principal constituents of the so-called fossil fuels, coal, petroleum, and natural gas.

Hydrocarbons are organic chemical compounds composed only of the elements carbon and hydrogen. There are five main types of hydrocarbon: kerogen, a fine-grained, amorphous mass of organic matter; asphalt, which is produced by the partial maturation of kerogen or the degradation of crude oil and is a solid at surface temperatures; crude oil, a mixture of thousands of different hydrocarbon compounds that is a liquid at surface temperatures; natural gas, which can be one of a number of different gases, including methane, ethane, propane, and butane, or a mixture thereof; and condensates, which are transitional between gas and crude oil.

These hydrocarbons form when large amounts of organic matter, which is rich in carbon and hydrogen, accumulates but doesn't decompose. This can occur in a number of different ways: for example, dead marine organisms, typically single-celled planktonic plants and animals, settle in a basin on the seabed at depths where the oxygen concentration is too low for decomposition to take place; in river deltas, sediment covers organic matter faster than decomposition can take place; or in swamp forests, vegetation falls into oxygen-depleted water.

As layers of fine-grained sediment are deposited on top

Below *An open-cast coal mine in Kemerovo Oblast in Siberia, Russia. The mine lies within the Kuznetsk Basin, one of the world's largest active coalfields, covering about 10,000 square miles (26,000 square kilometers) and with minable reserves of more than 300 billion tons (270 billion tonnes).*

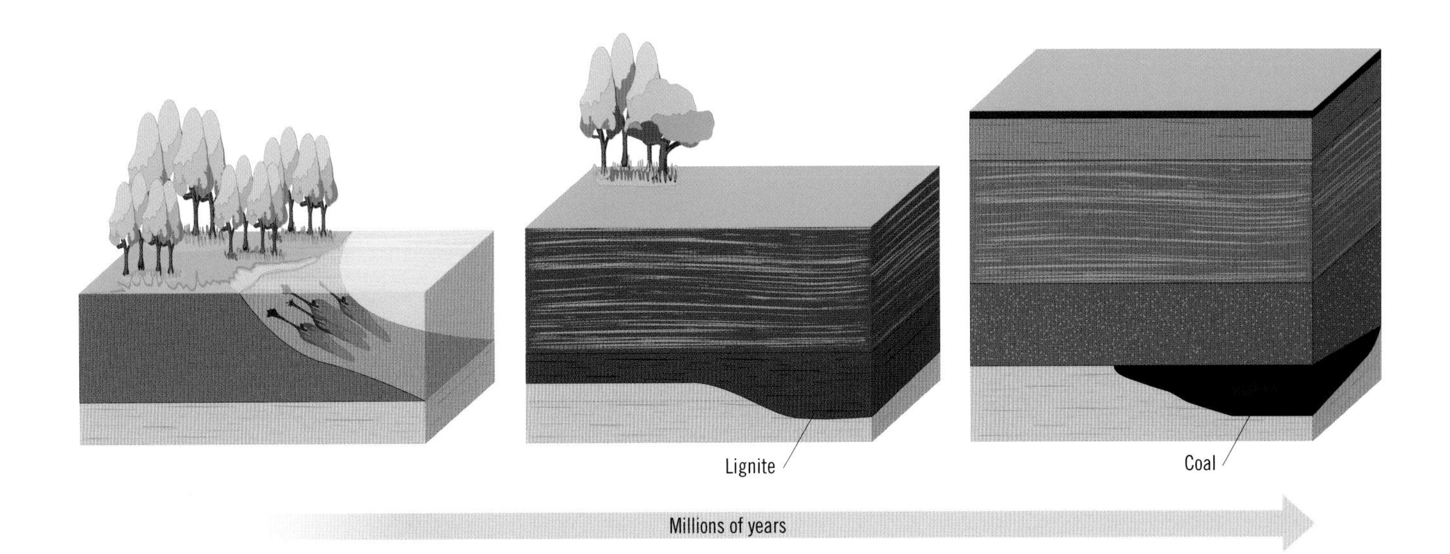

Below *Coal formation. Trees and other vegetation fall into a swamp, where anoxic conditions slow decomposition. Sediment eventually covers this material and, over millions of years, the weight of the overlying sediment and geothermal heat transform it into lignite and then coal.*

of the organic material, the weight of the overlying deposits pushes the material below the surface. The deeper it goes, the hotter it becomes, at a rate of about 1.6°F per 100 feet (3°C per 100 meters) of depth, and the gradual increase of pressure and temperature causes the transformation of the organic matter into hydrocarbons. Hence hydrocarbons are only formed in sedimentary rocks—almost always shales.

In the formation of oil and natural gas, the first stage in this transformation is the formation of kerogen. As the temperature and pressure in the source rock increase further, the kerogen is converted first into petroleum and then into natural gas. Depending on the amount and type of organic matter, hydrocarbon generation takes place at depths of about 2,450–16,400 feet (750–5,000 meters) and at temperatures of 140–300°F (60–150°C), an environment known as the "oil window." Above 300°F (150°C) and deeper than about 13,000 feet (4,000 meters) is the "gas window." The depth at which maximum hydrocarbon generation occurs is between 6,600 and 9,500 feet (2,000 and 2,900 meters).

Once this maturation phase has ended, the hydrocarbons enter the migration phase, in which they move from the impermeable source rock into the porous reservoir rock, typically sandstone or carbonates such as limestone. As the pressure increases, the lighter parts (oil and gas) are pushed out and move up through faults and pore spaces in the overlying rock until they are trapped in the reservoir rock by an impermeable layer known as a cap or seal.

Coal is made from the remains of prehistoric plants, typically from swampy areas. When the plants died, they fell into the water and were protected from decomposition by the lack of oxygen in the water. The woody parts of the plants accumulated and eventually became peat. Sediment washed over the peat and as new layers of sediment accumulated, the peat was compressed, squeezing out the water. The buried peat was then transformed by increasing temperature and pressure, first into lignite, then into bituminous coal, and finally into anthracite. In general, the deeper coal is found, the higher its quality, as the increasing heat and pressure force out the remaining water and other compounds, increasing the coal's hardness, density, carbon concentration, and energy potential. Although coal has been produced during most geological periods, 90 percent of the known coal beds were deposited during the Carboniferous and Permian periods.

Because they are rich in energy, hydrocarbons make excellent fuels. However, the burning of these fossil fuels causes the release of carbon dioxide, the main greenhouse gas that is causing global warming.

GAS HYDRATES

As well as forming by the thermal decomposition of organic matter, methane is produced by microbes as a product of their metabolism, especially during digestion. In some areas, particularly marine sediments but also in permafrost and some lake sediments, where temperatures are low and pressures are high, this methane and the surrounding water freeze to form an ice-like compound called a gas hydrate.

// Soils

Vital to life on Earth, soils are diverse and ever-changing.

Soil is composed of four ingredients: inorganic minerals (40–45 percent of soil volume), organic matter or humus (about 5 percent), air (about 25 percent), and water (about 25 percent). It also contains a significant community of microorganisms—a teaspoon of rich soil may host as many as a billion bacteria. The fact that solid matter only makes up about half of typical soils allows for the infiltration, movement, and retention of air and water, both of which are critical for life in soil.

Soil minerals comprise (from smallest to largest) clay, silt, and sand. Their relative proportions determine soil texture. There is a vast diversity of soil types; classification systems may contain several thousand. Soils without any one dominant particle size, containing a mixture of sand, silt, and humus, are called loams.

Formation of soil, or pedogenesis, is influenced by five main factors: parent material, climate, topography, organisms, and time. Soils are primarily the product of weathering and erosion, which break bedrock down into smaller and smaller pieces. In much of the Northern Hemisphere, they are derived from glacial erosion during the last glacial period. Soils are constantly changing as minerals and rocks weather, nutrients leach, and plant communities change. The mineral most commonly found in soils is quartz.

A soil's minerals may be derived directly from the weathering of underlying bedrock or may have originated elsewhere and been transported by wind or water. Deposits of windblown (aeolian) sediment are known as loess. They usually cover areas of hundreds of square miles in a thick blanket of fine-grained material and tend to develop into extremely rich soils; however, they are geologically unstable and erode readily.

The size of the particles in a soil largely determines its ability to retain water, clay and silt being more water-retentive, physically holding it through capillary forces. These forces are

Right *The mineral particles found in soils are categorized according to their size. From smallest to largest, the three types are clay, silt, and sand. The relative proportions of the different types are what determines a soil's texture.*

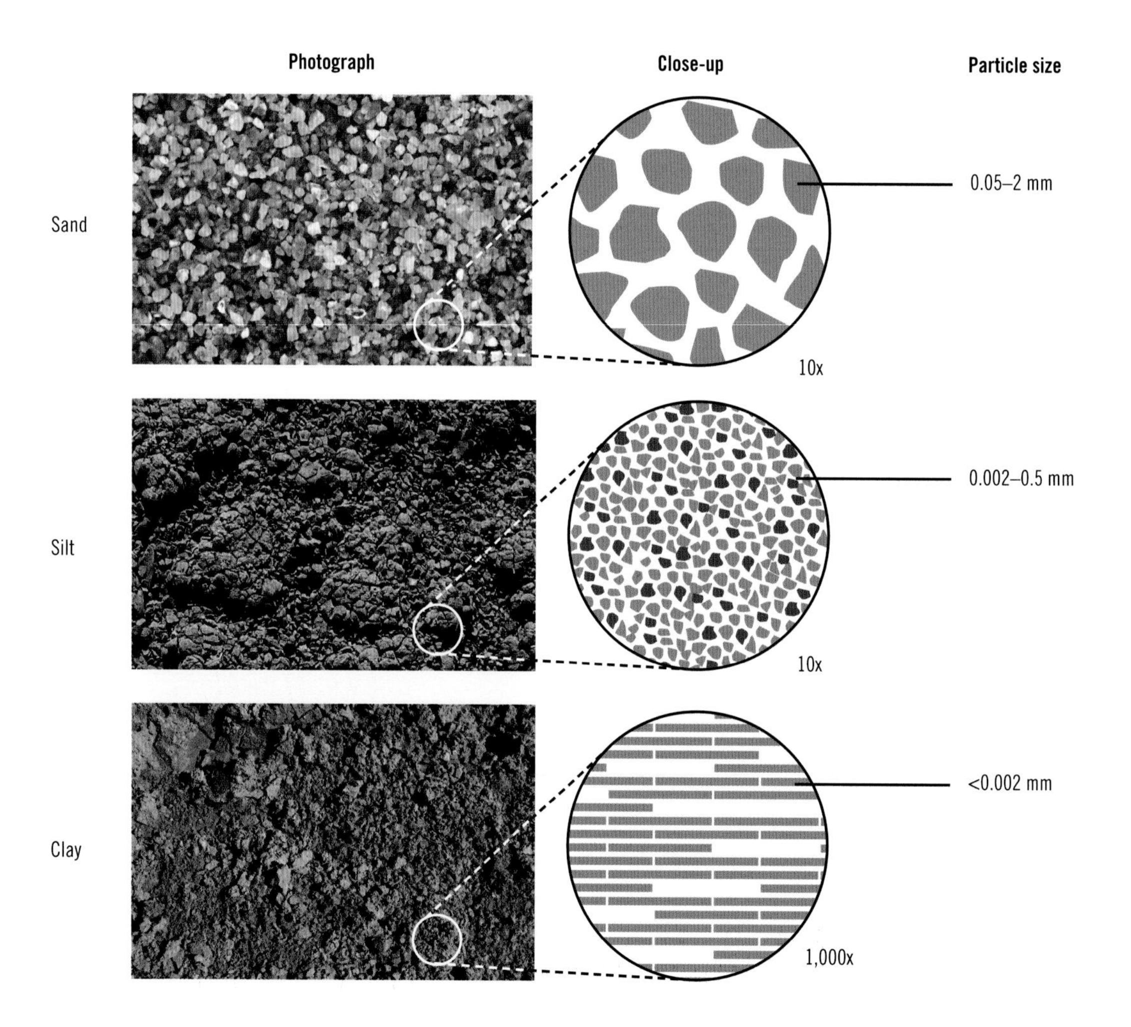

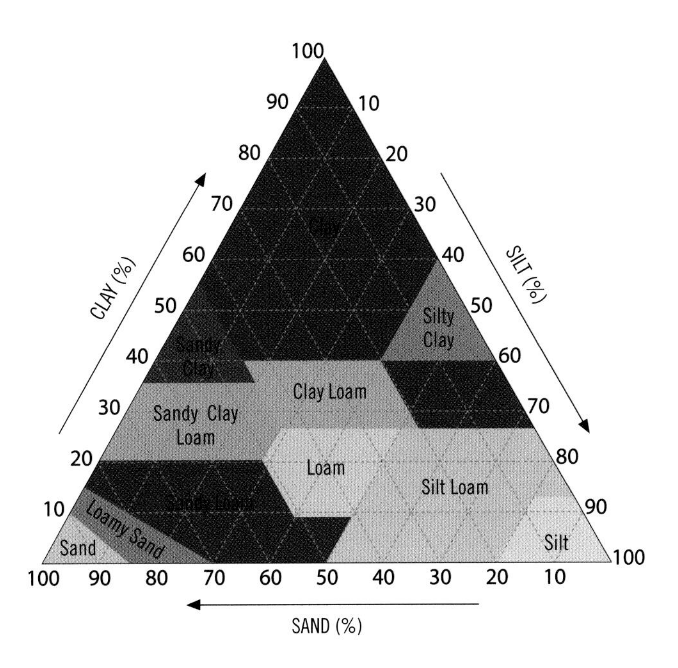

Above *Soil types are defined by the relative proportions of clay, silt and sand that they contain. Those in which there is no single dominant particle size are known as loams.*

strongest in clay, and can make it difficult to prise water away. Hence silt particles are best for plants.

Humus consists of living, dead, and decaying microorganisms, plants and animals. The presence of organic matter can greatly improve a soil's ability to hold water and the percentage of humus in a soil is among the best indicators of its suitability for agriculture. Humus is also an important source of nutrients.

Soils provide important ecosystem services, including recycling of nutrients, purification of water, and exchange of gases with the atmosphere. They are vital for the support of terrestrial ecosystems: they retain moisture and make it available to plants; cycle and recycle nutrients and make them available to plants; provide a habitat for organisms; regulate water quality, filtering and remediating pollutants; and reduce flooding by transferring water slowly to streams and groundwater. Soil also plays an important role in the Earth's carbon cycle, acting as a significant carbon reservoir.

Many of these services can be lost when soils are degraded or removed. Human activities can degrade soils by increasing the rate of erosion of the fertile topsoil through land clearing and overgrazing, or by contaminating it with salt through improper irrigation.

Water is critical to soil development, eroding, dissolving, transporting, and depositing minerals, organic matter and nutrients. The liquid found in the soil, known as the soil solution, is a mixture of water and dissolved or suspended organic and inorganic materials.

SOIL PROFILES

Soils typically have a number of layers, called horizons, with distinct physical and chemical properties. Their vertical arrangement is known as the soil profile. Horizons generally lack sharply defined boundaries, interacting with each other, and can vary significantly in thickness. Surface horizons tend to be dynamic and rich in life and organic matter, while lower horizons are more stable. Water percolating the soil and the activities of plant roots and animals such as earthworms can move material from upper layers into deeper layers.

Soil profiles are typically divided into four horizons:

O: Topsoil. Rich in humus (hence O for organic), which enriches the soil with nutrients and enhances moisture retention, along with microbes, which decompose the humus, and plant roots, which take up water and nutrients.

A: The beginning of true mineral soil. A mixture of organic material and inorganic weathering products. Typically dark-colored due to the presence of organic matter.

B: The subsoil. Mostly composed of fine material that has migrated down. Dense. May contain nodules or a layer of calcium carbonate.

C: The soil base. Lies atop the bedrock. Little affected by soil-formation processes.

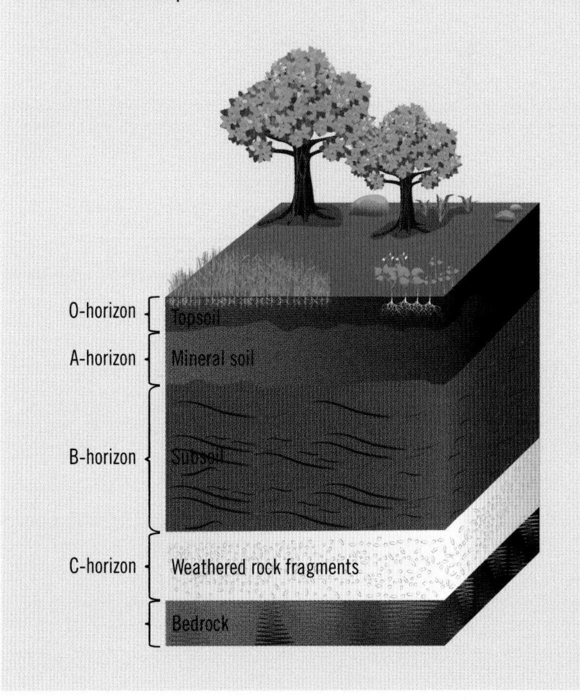

// Karst

When the underlying rock in a region is water soluble, the resulting landscape, known as karst, can be striking and distinctive.

Karst landscapes are shaped by the dissolution of rock along cracks and lines of weakness. The most common rock type in karst is limestone, but it can also form on gypsum and dolomite. Rain contains dissolved carbon dioxide, which makes rainwater weakly acidic. When it passes through soil, it often takes up even more carbon dioxide, forming a weak carbonic acid solution. Upon reaching limestone, the water begins to dissolve the calcium carbonate rock. As it seeps into cracks, it slowly widens them, allowing more water in and dissolving more rock.

Above *The area around Guilin in southern China was once a huge block of Devonian limestone. After being uplifted and exposed as a consequence of the collision of India with Asia, the bedrock has slowly been eroded away to leave the region's distinctive conical limestone hills and towers.*

Eventually, an underground drainage system begins to develop, increasing the rate of erosion even further.

Above the surface, karst can take many forms, depending on the nature of the underlying rock, from the steep-sided, round-summited hills of southern China and northern Vietnam to the razor-edged ridges of rock of the tsingy landscape in Madagascar. In regions where glacial action has exposed limestone bedrock, flat limestone pavements may form, with stone slabs called clints separated by deep fissures called grykes. Along limestone shores, particularly in the tropics, erosion by the ocean and marine organisms produces a distinctive landscape of sheer, cave-riddled cliffs and bays filled with vegetation-topped limestone pillars.

Below the surface, the erosional action of the water seeping through limestone can eventually lead to the formation of extensive cave systems. Indeed, most of the world's large cave areas are found in karst. If a cave extends close enough to the surface, its roof may collapse to form a depression known as a sinkhole, one of the most characteristic features of karst landscapes. Although some sinkholes develop gradually as an opening on the surface enlarges, in other cases the erosion is hidden beneath the surface. When the roof of the cavern suddenly collapses, cattle, cars, and even family homes can fall into the newly formed sinkhole.

Surface water and waterways are rare in karst areas because rainwater quickly drains away through cracks, crevices and sinkholes, often then forming underground streams and rivers. Sometimes these underground waterways become visible when the roof of an overlying cavern collapses. The hole through which the stream or river can be seen is known as a karst fenster or karst window.

Underground water systems are responsible for many unusual karst landforms. For example, each year, in some Irish karst areas, water wells up from underground water systems to form a unique type of seasonal lake called a turlough. Rivers in karst regions will often disappear into a sinkhole and then flow along the top of the water table in an underground cave system before resurfacing as a karst spring.

The global distribution of karst essentially maps on to the distribution of carbonate rocks, so karst landscapes are mostly found in sedimentary basins. Although karst landscapes make up about 13 percent of the world's land surface, they support about a quarter of its population. Many of these people depend on karst aquifers for their water supply. For example, the Syrian capital, Damascus, a city of more than two million people, draws virtually all of its water from karst aquifers.

Australia's Nullarbor Plain is the world's largest limestone karst. Occupying an area of about 77,000 square miles (200,000 square kilometers), it was once the bed of a shallow sea and today contains numerous caves, several of which are important archaeological sites.

Below *Western Madagascar's tsingy karsts formed through a mixture of horizontal erosion by groundwater and vertical erosion by rainwater. The name means 'where one cannot walk barefoot' in Malagasy.*

// Caves

Natural voids in the ground large enough for a human to enter, caves can extend deep underground and be adorned with mineral decorations.

Most caves form in karst, when slightly acidic water dissolves limestone (calcium carbonate), dolomite, or gypsum (or occasionally salt) to create subterranean caverns and passages. These are known as solutional caves. It typically takes more than 100,000 years for a solutional cave to become large enough to hold a person.

The distinctive rock formations found within limestone caves are called speleothems. They form as some of the carbon dioxide in the acidic water escapes, causing some of the dissolved calcium carbonate (also known as calcite) to come out of solution and deposit on the inside of the cave. The best-known speleothems are stalactites, which form as water drips from the cave ceiling, and stalagmites, which rise up from the cave floor, usually as a result of water dripping off a stalactite. When these eventually join together, they form a column. Flowing water may leave behind a sheet of calcite that eventually builds up into a deposit known as a flowstone. The colors of these cave decorations are the result of iron oxides or what are known as humic substances, which are derived from organic materials in the overlying soils.

Caves formed away from karst regions include glacier caves, long tunnels found near the snouts of glaciers between the glacial ice and the underlying bedrock, formed by meltwater flowing through and under the ice; sea caves, which form when waves widen fractures and other weaknesses in the bedrock of sea cliffs; aeolian caves, shallow caves often found in desert areas where wind scours chambers out of cliffs; talus caves, formed when boulders pile up on mountain slopes; and rock shelters, formed when erosion or weathering causes the removal of weaker rock, leaving a cavern of a more resistant rock.

Caves can also form when tectonic forces pull the bedrock apart. These tectonic caves tend to form in massive, brittle

Below *Cave formations, known as speleothems, take many different forms. Among the more common are stalactites and stalagmites, columns, straws, flowstone, and draperies.*

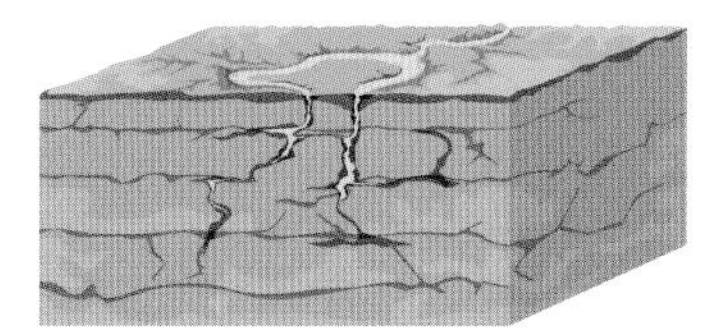

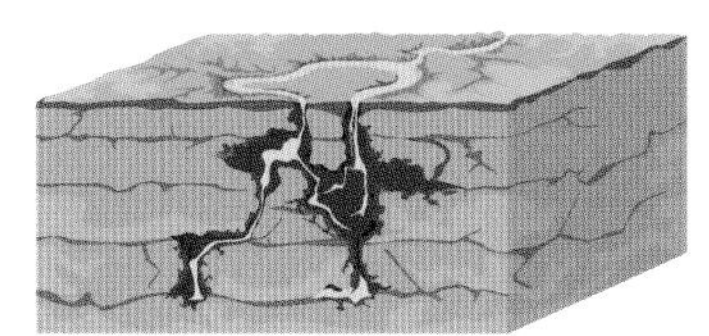

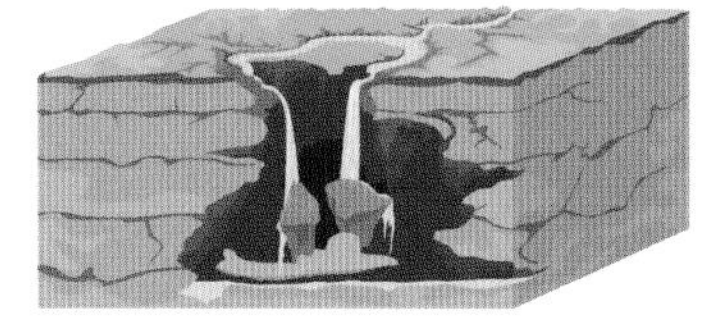

Above *Solution cave formation. Solution caves form when slightly acidic water seeps through cracks and dissolves the surrounding rock (usually limestone), widening the fissures. Underground streams eventually form, further eroding the rock and creating large caverns.*

rocks such as sandstone and granite, and take the form of high, narrow fissures with flat ceilings. Although they are among the most common caves, they are usually relatively small and are hence rarely noticed or cataloged.

Caves are found throughout the world; however, our knowledge of their abundance and distribution within a particular region is largely dependent on the popularity of caving in that region. The world's deepest known cave—measured from its highest entrance to its lowest point—is the 7,230-feet-deep (2,204 meters) Veryovkina Cave in Abkhazia, Georgia. The largest known cave by volume is the Clearwater cave system in Malaysian Borneo, which has a volume of roughly 134 million cubic feet (3.8 million cubic meters). The longest known cave is the Mammoth Cave-Flint Ridge system in Kentucky, which has a surveyed length of more than 345 miles (555 kilometers).

Many cave systems support simple animal communities. Creatures that live their whole lives in caves typically share a range of characteristics, including the loss of pigment and eyes, an elongation of appendages and an enhancement of other senses, such as the ability to detect vibrations in water.

Right *Lang Cave, Gunung Mulu National Park, Sarawak, Malaysian Borneo. Gunung Mulu is home to at least 183 miles (295 kilometers) of explored caves, which boast the world's largest underground chamber (the Sarawak Chamber), the world's largest cave passage (Deer Cave), and the longest cave in Southeast Asia (Clearwater Cave).*

LAVA TUBES

Volcanic activity can also lead to the creation of caves. Lava tubes form when highly fluid lava, notably a basaltic type known as pahoehoe (see page 22), flows downhill in a gully or other natural channel. The outer surface cools and solidifies, and if the source of the lava dries up, the long tube of solid lava empties out, leaving a near-cylindrical cave. Lava tubes can be tens of miles long and can even have lava stalactites and stalagmites, formed as molten rock drips from the ceiling to the floor. Huge caves can also form when a fissure opens up beneath a subterranean magma chamber and all of the molten rock drains away, leaving a large empty space.

The 443-foot (135-meter) long Thurston Lava Tube in Hawaii Volcanoes National Park, Hawaii, which formed during an eruption of Kīlauea more than 500 years ago.

// Landslides and avalanches

When a sloping area of land becomes unstable, it can suddenly give way in a potentially devastating landslide or avalanche.

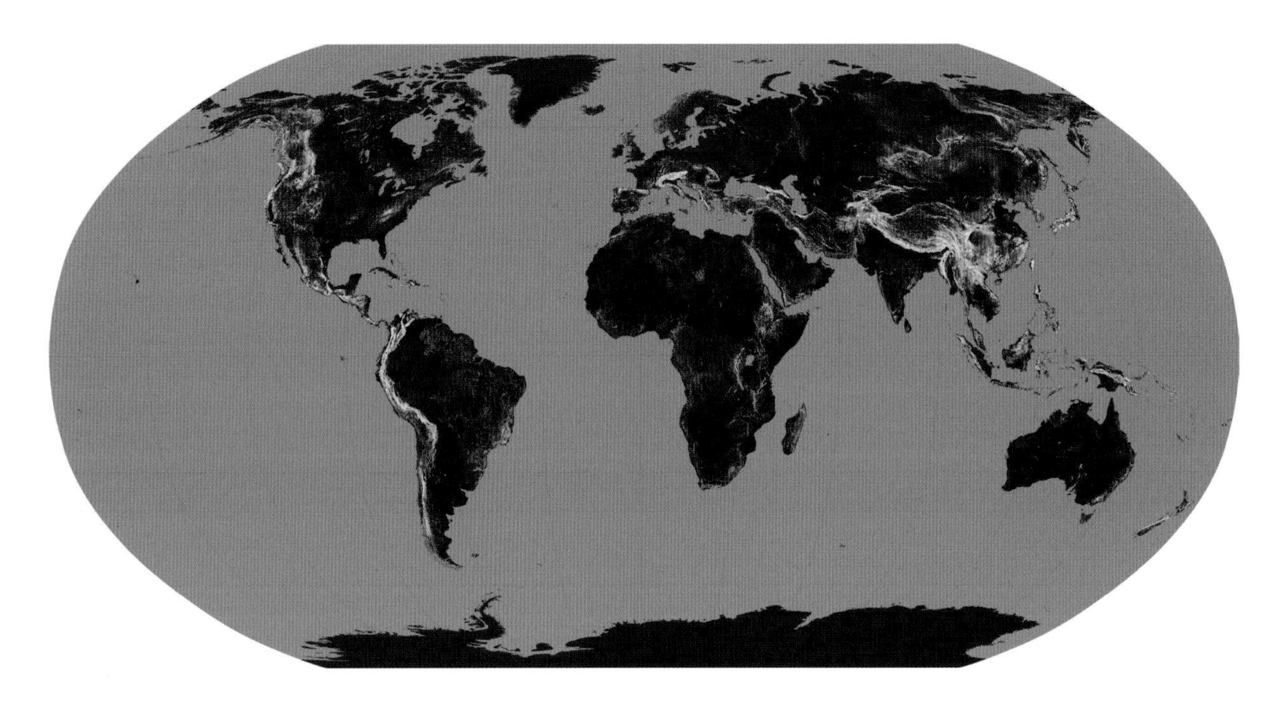

Above *Global landslide susceptibility, 2000–13. The most important factor that makes a particular landscape more susceptible to landslides is the steepness of the slope, but deforestation, the presence of roads, the strength of the bedrock and soils, and the location of faults are also key factors. Many of the areas that are most susceptible to landslides are found in steep mountain ranges.*

Landslides typically occur when the forces holding a slope together become weaker than the force of gravity pushing it downhill. But, while gravity is the ultimate driving force, several other factors, including heavy rainfall, slope steepness, an earthquake, and human activity, such as road building and the clearing of vegetation, increase the likelihood that a slope will give way.

Steepness is a key factor, as is the type of rock and the orientation of the rock layers. Unvegetated slopes are more vulnerable to landslides: the root systems of trees, bushes, and other plants help to bind soil together.

Among the most common causes of landslides is infiltration by water, whether from rainfall, snowmelt, changes in the depth of the water table, or some other source. Water adds weight, weakens the material, and reduces friction both among particles and with the underlying material, making it easier for material to move.

When conditions are in place for a landslide, all that is needed is a trigger—often a vibration, either natural, such as an earthquake, or man-made, such as construction, blasting, or mining. Erosion can also trigger a landslide: when the lower part of a slope is eroded away by the sea or a river, the slope eventually becomes too steep to stay up.

Landslides are typically classified according to the type of movement. The five main categories are falls, topples, spreads, slides, and flows. Landslides are complex, however, and more than one type of movement will often occur in a single landslide. They are also divided up according to the type of material involved: rock, debris, or earth.

In falls and topples, heavy chunks of material such as boulders detach from a cliff or a steep slope. A spread occurs when a relatively coherent slab of earth or other material moves, usually quite slowly (although we tend to think of a landslide as a rapidly moving mass of material, some creep along at just a few fractions of an inch per year). A slide takes place when there is a distinct zone of weakness that separates the sliding material from more stable underlying material. When the material moves as if it were a viscous fluid—usually due to the presence of water, but also sometimes because of trapped air—it's known as a flow. The more water there is in a flow, the faster it will move.

Volcanic landslides, also called lahars, are among the most devastating landslides. Resembling a churning slurry of wet concrete, they may contain hot volcanic ash, toxic gases, and lava, as well as debris. They can move at speeds of 125 mph (200 km/h) and may grow to more than ten times their initial size as they proceed downhill. The eruption of Mount St Helens in the US state of Washington in 1980 led to the largest landslide in recorded history, when a flow volume of about 0.7 cubic miles (2.9 cubic kilometers) spread out over an area of 24 square miles (62 square kilometers).

When sediment collapses underwater it's known as a submarine landslide, and can cause a tsunami. Tsunamis may also occur when large terrestrial landslides deposit material into the ocean.

Avalanches

Landslides involving snow are known as avalanches. They can range in size from small shifts of loose snow (known as sluffing) to enormous flows containing millions of tonnes of snow that can wipe out forests and destroy whole villages. Each year, about 150 people are killed by avalanches.

As snow falls, it builds up into what is known as the snowpack. When new snow falls, it creates a new layer of snowpack. If the layers don't bond well with each other, this can lead to a slab of snow, potentially thousands of cubic metres in size, detaching and beginning to flow downhill. This is a slab avalanche.

Once the slab begins to move, it usually accelerates rapidly, potentially reaching up to 80 mph (130 km/h). The avalanche grows quickly as it pulls in more snow and, as in a landslide, it will pick up other material, including trees and rocks, as it moves.

In rapidly moving avalanches, some of the snow may mix with air, forming what is known as a powder-snow avalanche. These can move at speeds of more than 185 mph (300 km/h) and may contain as much as 11 million tons (10 million tonnes) of snow.

Below *An avalanche falls from Khan Tengri, a mountain located in the central Tian Shan range, where the borders of Kazakhstan, Kyrgyzstan, and China meet.*

Below *Landslides can take several different forms, depending on the type of material involved and the steepness of the slope.*

Below *Most avalanches take place on slopes of between 25° and 45°. They occur when a slab of snow that hasn't bonded well with the snow below detaches and begins to flow down the slope. If the snow is relatively dry, it may mix with the air to form a powder snow avalanche.*

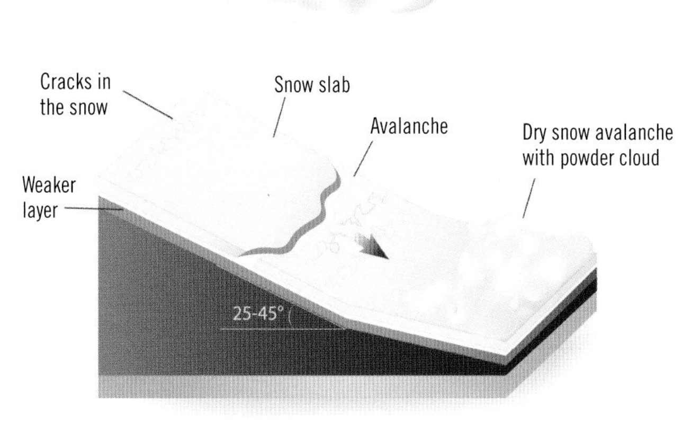

// Coastal landforms

Where land meets sea, the power of waves, tides, and currents sculpts the coastline into dramatic landforms.

The coast is where ocean energy is absorbed by land. As pounding waves, rising and falling tides and strong currents interact with each other and the rocks and sediments of the shoreline, they shape the land. The types of landforms found in these environments is largely dependent on whether they are dominated by forces of erosion or deposition.

Erosional coasts

Erosional coasts tend to have little or no sediment, exposed bedrock, steep slopes, high relief, and rugged topography. They are mostly found on the leading edges of lithospheric plates, on active margins with narrow continental shelves. The composition of the bedrock has a significant effect on the rate of erosion and types of landforms created on rocky coasts.

Sea cliffs: The most widespread landforms of erosional coasts, sea cliffs can be a few metres or several hundred feet high. At the base, there is often a wave-cut notch; cliffs remain vertical as material above the notch collapses.

Wave-cut platforms: At the base of most sea cliffs is a flat rock surface at about mid-tide level, called a wave-cut platform. Formed as wave action cuts the cliff back, it may be a few feet or several hundred feet wide.

Sea caves: Shallow caves formed in sea cliffs as less-resistant bedrock is eroded by waves.

Sea stacks and sea arches: The former are remnants of headland cut off from the mainland, while the latter are still connected. Sea arches form as weaker bedrock is eroded, and may collapse to form a sea stack. Both are impermanent features that will eventually be destroyed.

Headlands and bays: Where alternating bands of hard and soft rock run at right angles to the coastline (known as discordant), the softer rock is eroded away to form bays, while the harder rock remains as headlands. Concordant coastlines, where rock runs parallel to the coast, have fewer bays and headlands.

Depositional coasts

Depositional shorelines are characterized by abundant sediment accumulation. They are dynamic environments, constantly changing—sometimes subtly, sometimes dramatically—as waves, wave-generated currents, and tides move sediment from place to place. They tend to have low relief and to be located on passive margins with a wide continental shelf and low wave energies. Such coasts often feature estuaries and lagoons, and may develop river deltas.

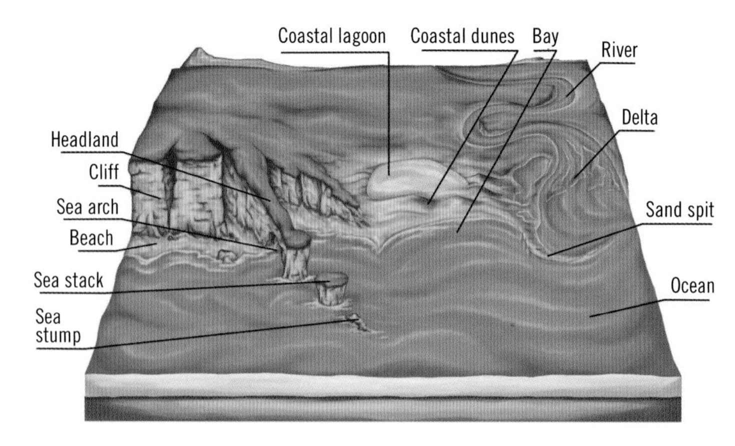

Above *The nature of the different types of coastal landforms is determined by a mixture of the bedrock type and the influence of the adjacent sea. They include erosional features such as sea stacks, sea arches and cliffs, and depositional features such as beaches, river deltas, and sand spits.*

Beaches: When a significant amount of sediment accumulates in one place, it forms a beach (see page 72).

Barrier islands: Long, narrow strips of sand formed primarily by a combination of waves and longshore currents, and generally oriented parallel to the shore. They stop waves and storms from reaching the shore, protecting fragile environments such as marshes, tidal flats, and lagoons on their landward side.

Barrier spits: Long ridges of sediment extending out from the coast. The growth of a spit can eventually cut a bay off from the ocean, creating a lagoon, in which case it is known as a bay barrier.

Tombolos: Spits that connect offshore landforms such as islands with the mainland. Tombolos form because refraction of waves around an offshore landform creates a zone of slow-moving water behind it, leading to deposition of sediment.

Coastal processes

The most significant processes for determining the nature of coastal landforms are waves, wave-generated currents, and tides.

Waves: Small waves move sediment towards the coast, depositing it on beaches; larger waves, especially during storms, move coastal sediment into deeper water. Severe storms can completely wash a beach away. Wave action erodes coastal bedrock largely through abrasion as sediment particles suspended in water are smashed against the shore. The force of waves smashing into the coast can also break up bedrock.

Longshore currents: Because waves usually approach at an angle, they are refracted as they enter shallow water. This creates a flow of water parallel to the shore, called a longshore current. Such currents extend from the shoreline out through the surf zone. Waves combine with longshore currents to move large quantities of sediment, which is swirled up from the sea floor by waves, then picked up by the longshore current. These currents can move in either direction along the coast, depending on the direction of wind and hence waves.

Rip currents: As waves roll in to a beach, there is a slight build-up of water at the shoreline. This causes water to move back out to sea along narrow pathways called rip currents, which sometimes carry sediment out.

Tides: Tidal currents can transport sediment and erode bedrock while also spreading wave energy and shifting the position of longshore currents by changing water depth and the position of the shoreline.

The Twelve Apostles are a series of sea stacks located off Australia's southern coast, formed as the relentless pounding of powerful waves from the Southern Ocean has eroded away the limestone cliff on the mainland.

// Beaches

Found in a wide range of forms, colors, and "textures," beaches are all shaped by a combination of water and weather.

Beaches are landforms made up of loose particles that are located beside a body of water, be it an ocean, lake, or river. The following mostly relates to coastal beaches.

In most cases, beaches are made up of a range of materials, including sand, pebbles, rocks, coral rubble, and seashell fragments. Sandy beaches are found in wave-dominated, depositional settings. Coastal beaches in the British Isles are often covered in rocks rather than sand: flat pebbles known as shingles and rounded rocks known as cobbles. Beaches near river mouths may be muddy, made up of finer-grained sediments that have been carried down in the river water.

Most of the material that makes up a beach is created by weathering and erosion, and will have come from a number of different sources. Some will have been transported from far away and some will be the result of the gradual breakdown of local features.

Beaches occur in a wide variety of colors, depending on the composition of their sand, which varies according to local minerals and geology. Most beach sand is yellow-orange quartz or silica, but beaches on volcanic islands can be black or green, while those on tropical islands may be composed of white fragments of coral.

The nature of the sediments found on a beach are an indication of the energy of the local waves and winds. Where coastlines are battered by strong winds and large waves, beaches will often be rocky rather than sandy. Highly protected coastlines may feature mud flats and mangrove forests as finer sediments are deposited.

Like other depositional landforms, beaches are highly dynamic, changing their shape and extent as they are sculpted by tides, currents and storms. During winter, storms may wash away sand that is then replaced by more gentle wave action in summer; hence, beaches are often narrower and steeper during winter. Tides tend to deposit sand as they rise and remove it as they fall.

A cross-section of a beach is known as its profile. Most beach profiles have four components: the swash zone—the area alternately covered and exposed by waves; the beach face—the sloping section that is exposed to the swash of the waves; the wrack line—the highest point reached by the swash zone; and the berm—the almost horizontal portion that stays dry except during extremely high tides and storms. The berm typically has a crest at the top. Right at the top, above high sea level, there may also be a flat beach terrace. And below the swash zone, there may be a trough and then one or more narrow sand embankments known as sandbars running parallel to the beach.

In some cases, sand dunes will develop behind the beach berm, as onshore winds carry beach sediment inland. The growth of vegetation can help to stabilize beach dunes; unvegetated dunes are more active and will change shape and position according to the weather.

The shape and dimensions of a beach depend on several factors, including wave type, tide height, and sediment composition and distribution. When higher-energy waves roll in and break in quick succession, sediment in the shallows is kicked up, making it susceptible to being carried along the beach by longshore currents or out to sea by tides; at these times, the beach will tend to have a gently sloping foreshore. During more quiescent periods, there is enough time between breaking waves for the water to recede and for the sediment to settle before the next wave arrives; at these times the beach will tend to have a steep foreshore. When sediments are fine, they will tend to compact, forming a smooth, gently sloping surface that is resistant to erosion by wind and water. Coarser-grained beaches typically have steeper foreshores.

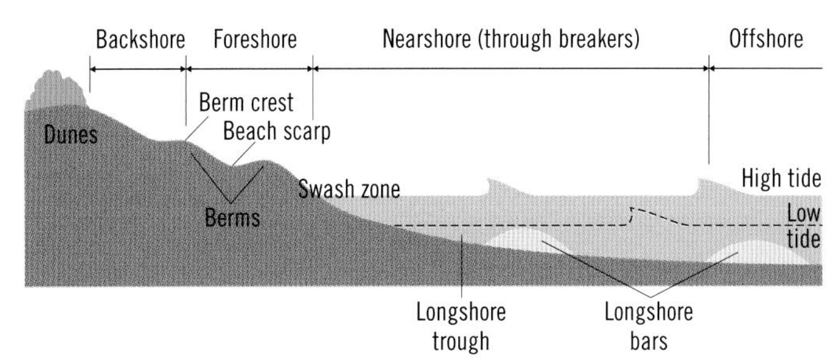

Above *A typical beach profile.*

Below *Beach and dunes, Jurien Bay, Western Australia. The sand dunes along this stretch of coastline are moving northwards at a rate of up to 52 feet (16 meters) per year.*

// Deserts

Barren, hostile landscapes where little precipitation falls, deserts are found on every continent, covering about a fifth of the Earth's land area.

Above *Sand dunes in the Sahara Desert, Morocco. With an area of 3.5 million square miles (9.2 million square kilometers)—almost a third of Africa—the Sahara is the world's largest hot desert. Sand seas make up only a small part of the desert's area, which mostly consists of rocky plateaus. For several hundred thousand years, the precession of the Earth's axis as it rotates around the Sun has caused the Sahara to switch between desert and savannah grassland every 20,000 years or so.*

Deserts are defined as areas that receive less than 10 inches (250 millimeters) of precipitation a year. As a result, most deserts have little vegetation. Water courses tend to be ephemeral, usually only filling after rain falls outside the region.

Although rainfall in a desert is rare, it is still the main source of water. When it does fall, it is usually localized and often heavy. Quick, violent rainstorms can have a significant effect on the landscape, with flash floods scouring surfaces and carving out deep channels known as arroyos or wadis. Where water collects in basins, it can form ephemeral lakes.

In some areas, deserts form near the leeward slopes of mountain ranges as a result of the rain-shadow effect. As moisture-laden air masses collide with mountain ranges, they are forced to rise. The air cools and drops the moisture it is carrying as rainfall on the mountains' windward slope. As it descends on the leeward side, it warms again, increasing its capacity to hold moisture and keeping the land in the mountains' rain shadow dry. Similarly, deserts at the center of continents, known as interior or inland deserts, and high-altitude montane deserts, form because air masses along the coast have released all of their moisture before they reach them.

In coastal regions where the water offshore is particularly cold, typically the western edges of continental land masses adjacent to upwellings, coastal deserts can form. Onshore winds pick up little moisture from the cold seawater, so rainfall is low; often the main form of precipitation is fog or dew. The driest place on Earth—the Atacama Desert—is a coastal desert; there are weather stations in the Atacama that have never recorded a drop of rain.

Subtropical deserts, found between 15° and 30° north and south of the equator, form because of the circulation patterns of air masses. As hot, moist air rises near the equator, it cools and releases its moisture in the form of heavy tropical rains. This leaves the air mass cooler and drier. It moves away from the equator and into the tropics, where it descends and warms up again, hindering cloud formation and reducing rainfall on the land below.

In addition to the well-known hot deserts, there are also cold deserts. These are regions, mostly found towards the poles but also at high altitudes, where minimal precipitation occurs and the air is so cold that it holds little moisture. Almost all of Antarctica is technically desert, making it the world's largest; on the central plateau, the annual

Above *Rain shadow desert formation. Warm, moisture-laden air rises as it hits a mountain range (red arrow) and cools (blue arrow), causing the water vapor to condense to form clouds. The clouds release the moisture in the form of precipitation, so the falling air on the other side of the mountains is dry (white arrow), creating an orographic or rain shadow in which there is little rainfall, causing arid desert conditions.*

precipitation is about 2 inches (50 millimeters). Some areas, including the McMurdo Dry Valleys, are ice-free due to the freezing, dry katabatic winds that flow downhill from the surrounding mountains and the scarcity of snowfall.

Hot deserts typically experience large daily and seasonal temperature changes. Clear skies by day mean that most of the Sun's radiation reaches the ground, raising temperatures to highs of more than 113°F/45°C (soil surface temperatures can be even higher, reaching 172°F/78°C). At night, heat is lost quickly as there is no humidity or clouds to provide insulation. The difference between night and day can be as much as 54°F (30°C). In winter, overnight temperatures can drop below freezing.

Below *Map of the world's deserts. Most of the Earth's hot deserts are found in two bands running between the latitudes of 30° and 50° north and south—the regions in which dry air moving from the high-altitudes toward the poles descends.*

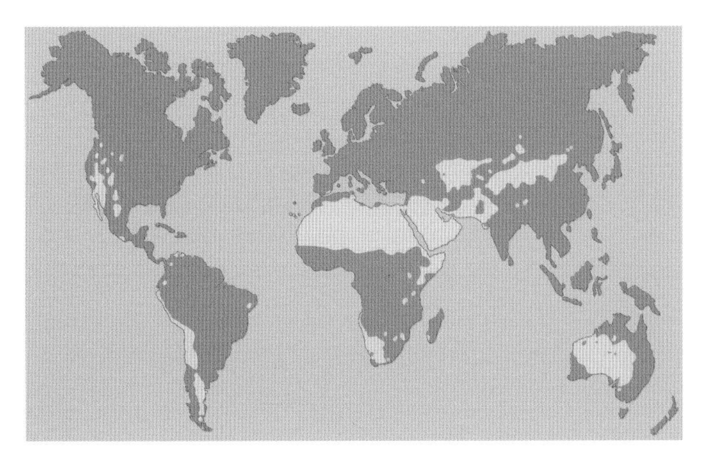

Above *The Valle de la Luna (Valley of the Moon) in the Atacama Desert in northern Chile. Located near the coast between two mountain chains (the Andes and the Chilean Coast Range), which create a double-sided rain shadow, the Atacama is the world's driest non-polar desert.*

The large variation in temperature can drive the weathering of desert rocks as the stress of constant thermal expansion and contraction causes them to crack, exfoliate, and shatter. Rain falling on hot rocks can have the same effect. Desert winds continue the erosion process, picking up grains of sand that abrade surfaces. In some areas, sand and dust are blown away to leave stony plains known as desert pavements, while in others, the sand is deposited in level areas termed sand fields or sand seas, or is piled up into dunes. About a fifth of the world's deserts are sandy, but sand dunes cover only about 10 percent of the world's deserts.

The shape of sand dunes is determined by the characteristics of the prevailing wind. When conditions are right, they can reach heights of 1,600 feet (500 meters). Although solitary dunes sometimes occur, more often they form dune fields, the largest of which are known as sand seas or ergs. The dunes themselves are constantly migrating, usually shifting by a few metres a year, although a violent sandstorm can move a dune by 65 feet (20 meters) in a single day.

About a sixth of the Earth's population—around a billion people—live in desert areas. In some cases, typically the semi-arid desert fringes, human activities such as overgrazing and poor irrigation practices cause deserts to grow. Global climate change's effect on temperatures and rainfall patterns is also likely to have an impact on this process. Each year, about 2.3 million square miles (6 million square kilometers) of land become useless for cultivation due to desertification.

The Sahara is the world's largest hot desert, with an area of 3.5 million square miles (9.2 million square kilometers).

Most deserts today are relatively young, geologically speaking, having developed during the Cenozoic Era (65.5 million years ago to the present), which has seen gradual cooling and consequent aridification of the global climate.

Where deserts sit atop aquifers, springs sometimes reach the surface, forming what are known as oases, where plants and animals can survive.

// Plains

Broad areas of relatively flat land, plains are one of the Earth's major landforms, covering a little more than a third of the terrestrial surface.

Present on every continent other than Antarctica, plains vary greatly in size, covering anywhere from a few hectares to hundreds of thousands of square miles. They can be found from north of the Arctic Circle (where they are known as tundra) to the tropics. While the largest plains are located in the interiors of the continents—most of central Australia, for example, is one vast desert plain—there are also coastal plains formed by the build-up of river sediments.

There are two main types of plain, defined by how they formed: depositional plains are created when rivers or glaciers deposit layers of sediment or when volcanic activity causes lava to flow out over the surface; erosional plains form when wind (pediplains), rivers, or glaciers (peneplains) scour away surface features.

Rivers that meander across relatively flat valley floors, dropping sediment as they go, can produce so-called scroll plains. When rivers repeatedly break their banks, which typically occurs on a seasonal basis, they often

An alluvial fan in the Taklamakan Desert, Xinjiang, China. Alluvial fans are relatively common in deserts, forming in areas where flash floods wash alluvium down from the adjacent hills.

Below *Formation of an alluvial fan. When water flowing in a confined channel such as a narrow canyon emerges from an escarpment, it spreads out in a fan shape and begins to seep into the surface, leaving behind the sediment and other material it was carrying, which is known as alluvium.*

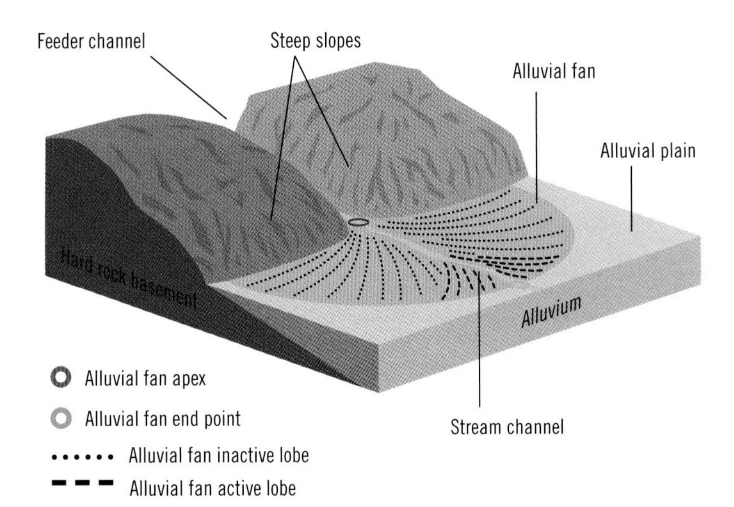

form floodplains, as their sediment loads are deposited in broad, flat alluvial fans. Similarly, a glacial outwash plain, also known as a sandur, is formed when layers of gravel and sand are deposited by meltwater at the terminus of a glacier. Sometimes, part of a glacier can become detached and melt, depositing the sediment that it was carrying and creating a so-called till plain. And when a lake dries up, the sediments at its bottom can form what is known as a lacustrine plain.

Plains support a wide range of habitats, from forest to desert to grasslands. These grasslands have different names, depending on where they occur: in Asia and eastern Europe, temperate grasslands are called steppes, while in North America, they are often called prairies; tropical grasslands are called savannahs.

Many of the world's plains are important sites for agriculture. Being flat, they are easily cultivated and their soils are typically deep and fertile. The grasslands that are found on many plains also offer good grazing for livestock.

Steppe landscape in Astrakhan, Russia. The semi-arid plains of the Eurasian or Great Steppe extend 4,971 miles (8,000 kilometers) from Hungary to China—almost a fifth of the way around the Earth—and support the world's largest temperate grassland.

// Wetlands

Often supporting highly productive and biodiverse ecosystems, wetlands are transition zones, neither totally dry nor totally under water, and are home to both aquatic and terrestrial species.

Wetlands are areas that are covered with water (fresh, salt, or brackish), or saturated with water, for at least part of the year. They are found on every continent, from the tundra to the tropics, and range in size from isolated prairie potholes to vast flooded forests. Among the world's largest wetlands are the Amazon River basin, the West Siberian Plain, the Pantanal in South America, and the Sundarbans in the Ganges-Brahmaputra delta in Bangladesh and India.

There are many forms of wetland, depending on their location, soils, topography, vegetation, water chemistry, and type of inundation (known as the hydrology). They can be permanent or ephemeral, flooding seasonally or irregularly following heavy rains. Their water may be groundwater seeping up from a spring, come from a nearby river or lake, or seawater that rises and falls with the tides. The level of water saturation is the main influence on the type of ecosystem that develops within a wetland.

Typically, wetlands are divided into coastal or tidal wetlands and inland or non-tidal wetlands. Often found in and around estuaries, coastal wetlands are dominated by plants that have adapted to survive in constantly varying salinities.

Wetlands are known by many different names, including peatlands, sloughs, muskegs, carrs, potholes, pocosins, quagmires, and vernal pools. However, the four main types are swamps, marshes, bogs, and fens.

Swamps are permanently saturated, tree-dominated wetlands most commonly found in the tropics. In coastal areas, mangroves are often the dominant tree type; the Sundarbans supports the world's largest mangrove forest.

In temperate regions, swamps give way to marshes, which are dominated by herbaceous plants such as grasses, rushes, or reeds, rather than woody species. They are typically found at the edges of lakes and streams, and at river mouths and along coastlines.

Bogs are wetlands that accumulate peat, a deposit of dead and decaying plant material, typically sphagnum moss. They are often covered in grasses, sedges, or shrubs and receive most of their water from precipitation. They are

The Pantanal, an enormous internal river delta, is the world's largest tropical wetland. Located mostly in Brazil but also extending into Bolivia and Paraguay, it covers an area of 54,000–75,000 square miles (140,000–195,000 square kilometers). Each year, during the rainy season, about 80 percent of the floodplain is submerged. This water is then slowly released into the Paraguay River and its tributaries.

usually found in colder regions, at either higher latitudes or higher altitudes, in areas where the surface water is acidic and nutrient-poor.

Fens, like bogs, have low nutrient availability and accumulations of peat. However, they are usually fed by surface water or groundwater and their water chemistry is either alkaline or has a neutral pH. Because they are typically nutrient-poor, bogs and fens often support communities of carnivorous plants.

Wetlands are extremely valuable ecosystems. They can play an important role in flood mitigation, absorbing excess water during heavy rainfall. Coastal wetlands help to protect and stabilize the shoreline, and often purify water by filtering out river pollutants. They are particularly important for migratory birds, offering a place to stop off to feed and rest. Many also support economically valuable fisheries and act as nurseries for juvenile ocean-going fish species.

However, wetlands have long been considered pest-ridden wastelands and, all over the world, many have been drained for development or flooded for recreational use or hydropower generation. At least half of the world's wetlands have been drained or otherwise destroyed. Indeed, wetlands are suffering more environmental degradation than any other ecosystem on Earth. Increasingly, city planners are creating artificial wetlands to serve as wastewater-treatment facilities or for flood control.

Above *A peat bog on the Isle of Skye, Scotland. Typically located in glacial rock depressions, Skye's bogs are believed to have begun to form at least 5,000 years ago.*

Below *Types of wetland. Marshes are dominated by herbaceous plants and are typically found at the edges of lakes and streams, at river mouths, and along coastlines. Bogs are wetlands that accumulate peat and receive most of their water from precipitation. Swamps are dominated by trees.*

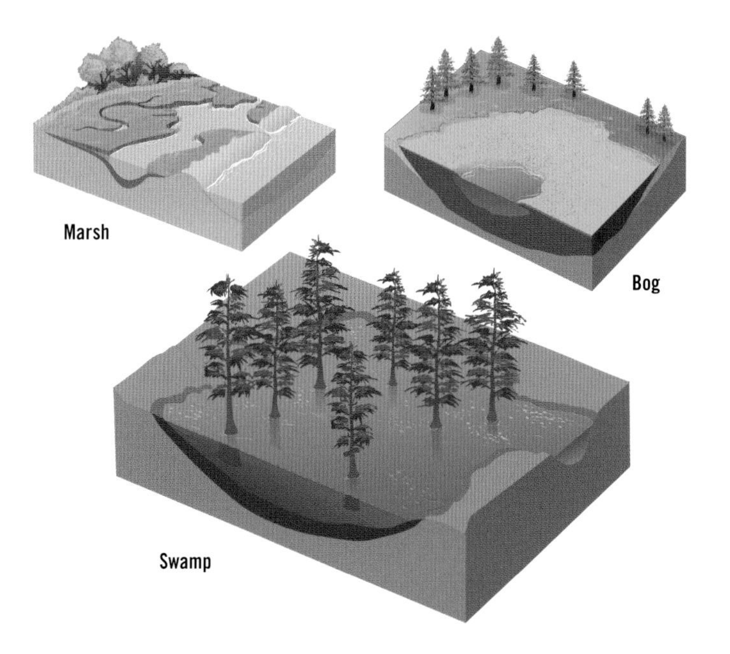

// The polar regions

The Earth's northern and southern extremes receive little energy from the sun and are among the coldest places on the planet.

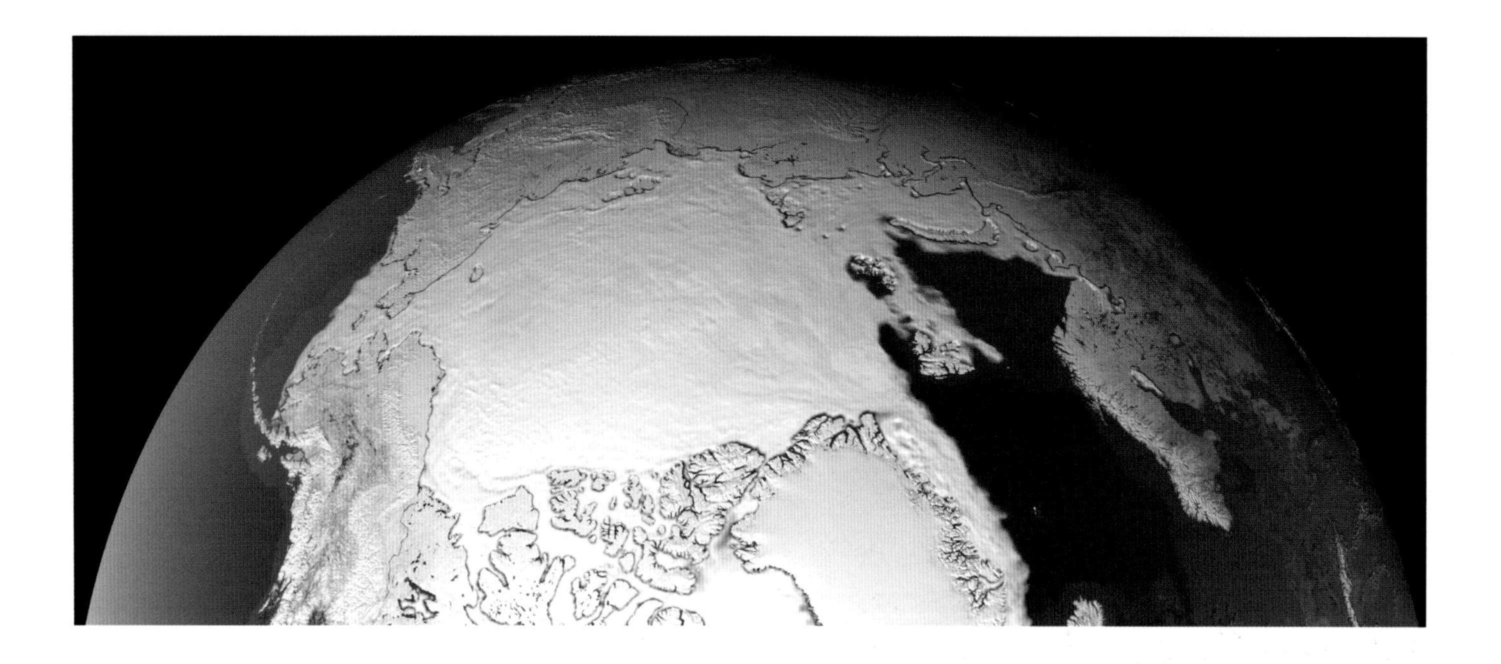

Above *The Arctic sea ice its annual maximum extent for 2021, which occurred on March 21.*

The extent of the polar regions is poorly defined. One measure is the Arctic and Antarctic Circles, at 66° 33' North and South—the latitudes at which the Sun does not set for exactly 24 hours during the summer solstice.

In polar regions, the Sun's energy arrives at an oblique angle, so it is spread over a larger area and must travel farther through the Earth's atmosphere, where it may be absorbed, scattered, or reflected. Hence solar radiation is less intense and temperatures colder than in other areas. Extensive snow and ice cover also reflect much of this weak sunlight, contributing to the cold. The oblique angle also ensures that there is extreme variation in daylight hours, from 24 hours of daylight in summer to complete darkness at mid-winter.

Differences in the geography of the two polar regions mean that they have different climates. The northern polar region, the Arctic, is centered on an ocean basin mostly surrounded by land, while the southern polar region, the Antarctic, is based on an ice-covered continental landmass.

The current situation, with both polar regions ice-covered simultaneously, is rare. The present state is partly due to the fact that the Earth is in the grip of an ice age (see page 40), but also because the continents are arranged in such a way as to cause patterns of air and ocean circulation that allow cold climatic conditions to prevail in both north and south.

About 98 percent of Antarctica's 5.5 million square miles (14.2 million square kilometers) is covered by ice up to 15,400 feet (4,700 meters) thick. The continent is also surrounded by floating ice shelves that extend out into the Southern Ocean. It is divided in two by a wide trench that is

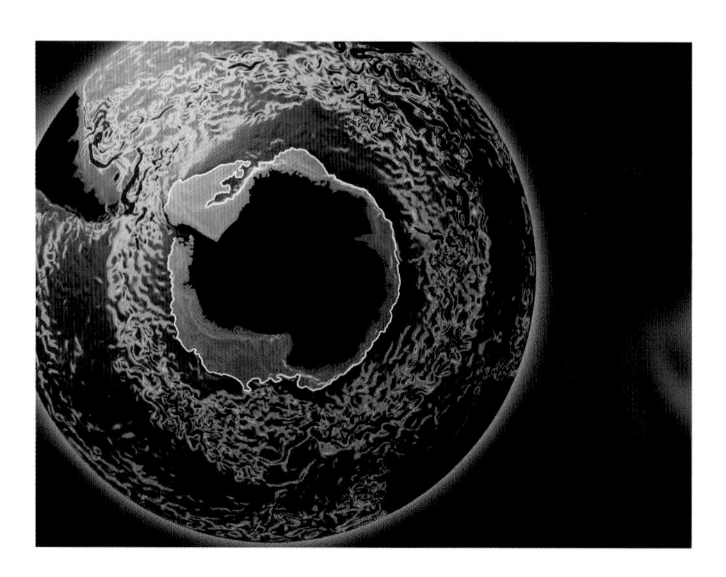

Above *The Antarctic Circumpolar Current is the world's largest ocean current. By keeping warmer water away from Antarctica, it allows the continent to maintain its ice sheet and ice shelves.*

Above *The Antarctic continent is divided into two sections (East and West Antarctica) by a wide trench.*

PERMAFROST

Permafrost—ground that remains completely frozen for two or more years—currently covers about 20 percent of the Earth's land surface. Although mostly in the Arctic and sub-Arctic, it also occurs at high elevations—including on Africa's highest peak, Kilimanjaro, about 3° south of the equator—in Antarctica and even under the sea.

In areas not overlain by ice, permafrost is covered in a layer of soil that thaws during warmer months. This "active" layer may be only 4 inches (10 centimeters) thick or may extend several metres. The surface layer of the permafrost itself contains large quantities of organic carbon derived from plant material that cannot decompose due to the cold. The permafrost then continues to the depth at which geothermal heat maintains a temperature above freezing. In parts of Siberia, that may be as much as 1 mile (1.5 kilometers) below the surface.

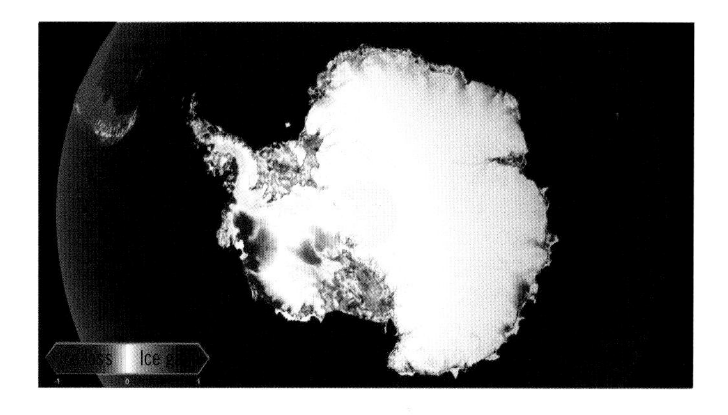

Above *Changes to the thickness of Antarctica's land ice between 2003–09 and 2018. While rising temperatures are causing ice to melt in some parts of the Antarctic, increased snowfall is causing the ice thickness in other areas to rise.*

widening by 3 feet (1 meter) every 500 years. On one side is East Antarctica, a large, 25-mile-thick (40-kilometers) slab of 3.8-billion-year-old continental crust; on the other is West Antarctica, made up of four smaller, thinner, loosely connected blocks of crust. A vast mountain range runs along the East Antarctic side of the trench for some 2,200 miles (3,500 kilometers), rising to more than 13,000 feet (4,000 meters). Much of West Antarctica is actually below sea level; without ice cover, it would be a group of islands.

The Antarctic is surrounded by the Southern Ocean and the Antarctic Circumpolar Current, insulating it climatically—one reason why it is generally much colder than the Arctic. The average annual temperature at the South Pole is -57°F (-49.3°C); at the North Pole, it is 0°F (-18°C).

Antarctica is a cold desert; much of the polar plateau receives less than 50 millimeters (2 inches) of precipitation annually. Around the continent's margin, regular cyclonic storms occur, dropping more than 300 millimeters (12 inches) of snow per year; however, strong katabatic winds blow much of it into the sea. Overall, mean annual snowfall for Antarctica is 6 inches (150 millimeters) or less. In contrast, on the Greenland ice sheet, the mean annual snow accumulation is about 15 inches (370 millimeters).

Temperatures on the polar plateau of East Antarctica are by far the coldest on Earth; at the Russian Vostok station, 11,500 feet (3,500 meters) above sea level, the mean annual temperature is -72°F (-58°C) and during the winter of 1983 it dropped to -128.5°F (-89.2°C). Coastal areas experience extremely strong katabatic winds; mean annual wind speed at Commonwealth Bay is 45 mph (72 km/h).

Left *Snow-covered mountains and glaciers on the Antarctic Peninsula. Like much of West Antarctica, the peninsula is actually a series of bedrock islands connected by an ice sheet.*

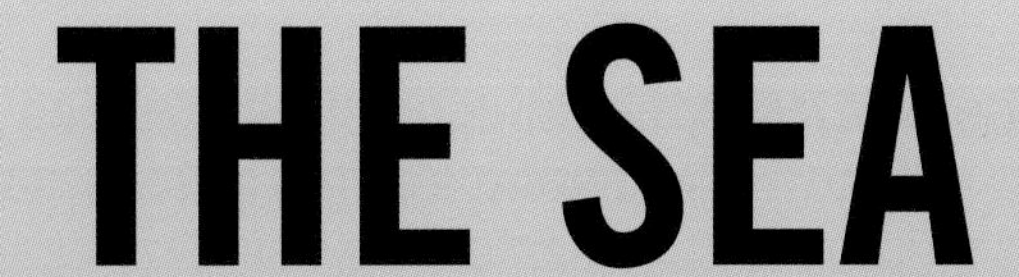

The Earth's surface is more sea than land, so it is hardly surprising that the oceans play a central role in many of the planet's most important processes. They regulate the world's temperature, absorbing more than half of the heat that reaches the Earth from the Sun and redistributing it around the planet, in the process influencing both global climate and local weather—generating powerful winds, destructive storms, deadly floods and devastating droughts. They play a central role in the water and carbon cycles. The oceans hold 97 percent of the Earth's water; most evaporation comes from the oceans; and most precipitation falls on the oceans. They are the cradle of life on Earth, today home to some 240,000 known species. They are the source of bounteous food: the oceans are the main provider of protein for more than a billion people and fish accounts for about 16 percent of the animal protein consumed globally. They are recreational playgrounds and the holder of the Earth's last great mysteries. They even affect the air we breathe: more than 70 percent of the oxygen we breathe is produced by marine plants. Sadly, they have also long been a dumping ground for human waste.

The sea is in constant motion and that restlessness shapes the world around us, from the waves that sculpt the coastlines to the ocean currents that help to determine regional climates.

// The oceans

The Earth's oceans play the defining role in its life-support system, regulating both heat and climate, generating oxygen and providing food for billions of people.

Roughly 71 percent of the Earth's surface is covered by ocean, an area of about 140 million square miles (362 million square kilometers). They contain 97 percent of the Earth's water, or about 324 million cubic miles (1.35 billion cubic kilometers) of water at an average depth of almost 12,000 feet (3,700 meters). The total mass of the oceans is around 1.54 quintillion (1.54×10^{18}) tons (1.4 quintillion tonnes), which represents about 0.023 percent of the Earth's total mass.

Approximately two thirds of the Earth's surface is covered by water more than 660 feet (200 meters) deep and nearly half of the oceans' waters are found below a depth of 9,800 feet (3,000 meters).

Although technically the Earth possesses a single ocean that encompasses the whole planet, geographers divide it up into five major oceans whose boundaries are largely defined by the continents that frame them. These are, in descending

Below *The ocean seen from space. Covering more than 70 percent of the Earth's surface, the oceans are effectively the planet's largest and most effective solar energy collector, absorbing large amounts of heat without increasing significantly in temperature.*

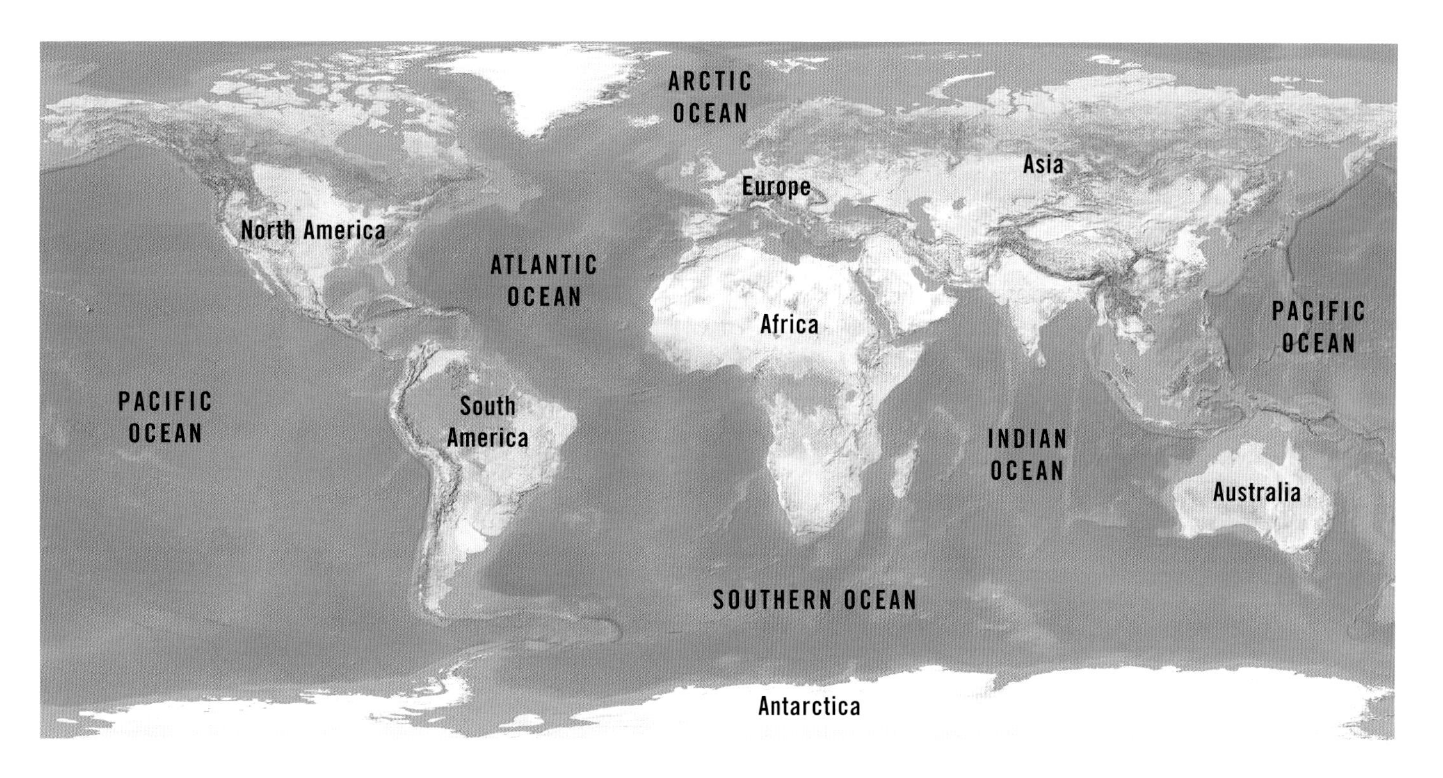

Above *Geographers divide the Earth's single globe-encompassing ocean into five major ocean basins whose boundaries are largely defined by the margins of the continents.*

order by area, the Pacific, Atlantic, Indian, Southern and Arctic oceans. Seas are smaller water bodies that are partly or fully enclosed by land.

Distribution of the water in the oceans is not spread equally over the Earth: 61 percent of the Northern Hemisphere's surface area is covered by oceans, compared to 81 percent in the Southern Hemisphere.

Color is a key characteristic of the ocean: its bluish tone is the result of the red end of the visible-light spectrum being absorbed by the water, leaving behind the colors at the blue end of the spectrum. Organic and inorganic particles in the water can change its hue, making it greener, as light bounces off them.

The origins of the global ocean, including when it first formed, are poorly understood. It is thought that almost all of the world's water was initially found in the chunks of space rock that joined together to form the Earth. This water emerged from molten rocks as water vapor in a process known as degassing or outgassing. Then, about four billion years ago, when the Earth cooled to below 100°C (210°F), the water vapor began to condense and fall as rain, before draining into hollows to create the primeval ocean.

The formation of the ocean was probably responsible for the emergence of life on Earth. Fossils of bacteria and cyanobacteria (blue-green algae) from the Precambrian Era—about 3.3 billion years ago—indicate that water was present at that time.

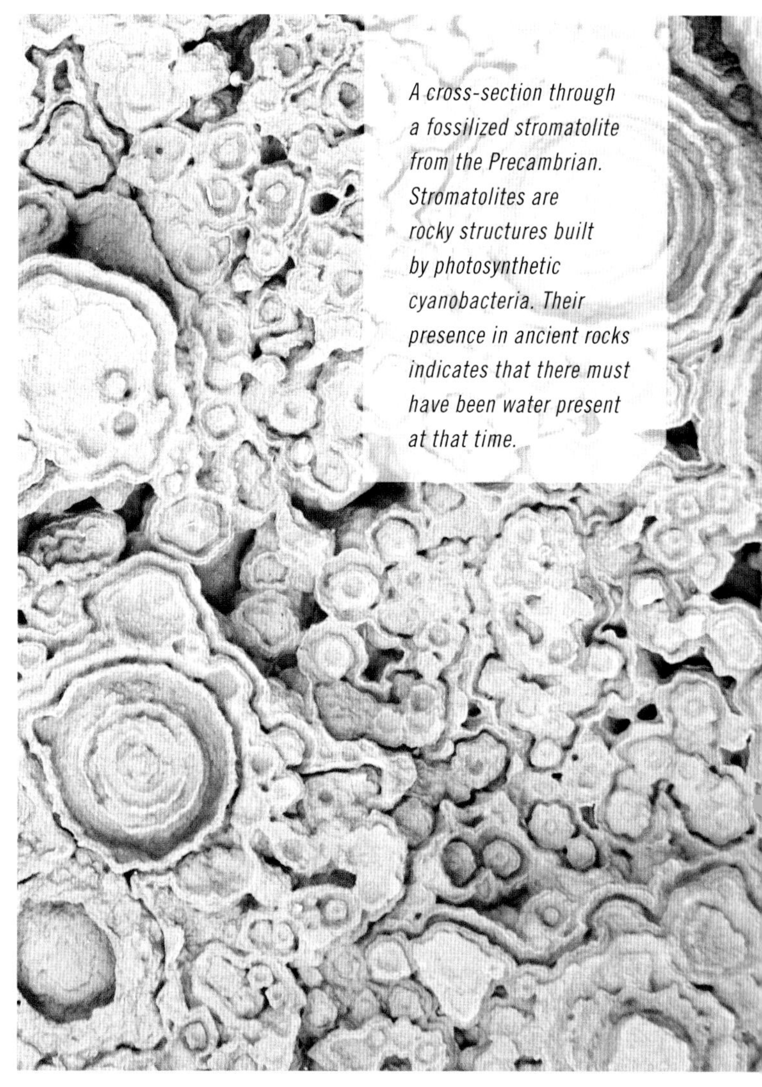

A cross-section through a fossilized stromatolite from the Precambrian. Stromatolites are rocky structures built by photosynthetic cyanobacteria. Their presence in ancient rocks indicates that there must have been water present at that time.

// Why is the sea salty?

Most of the Earth's surface is covered by salty water. But where did that salt come from?

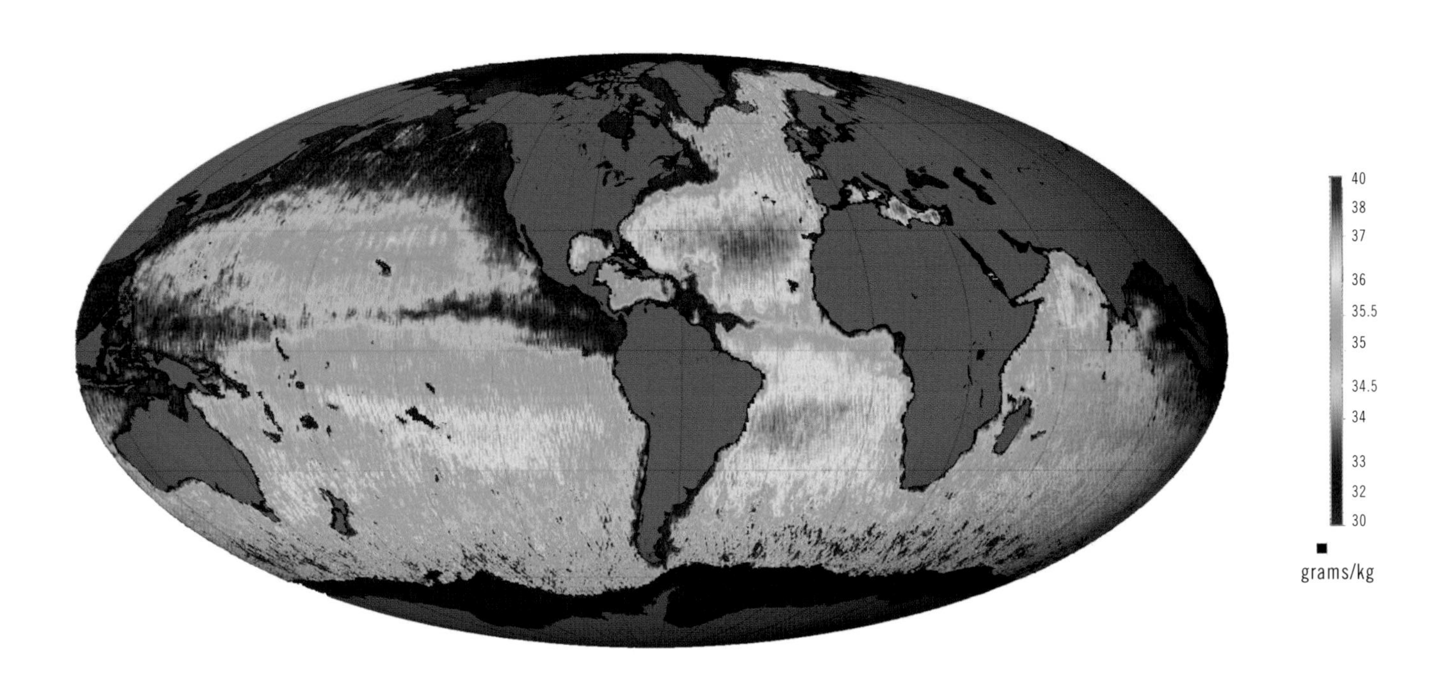

Above *Global ocean surface salinity map. As these satellite data were collected during the Northern Hemisphere summer, salinity is low in the Arctic due to the melting of sea ice. Conversely, high evaporation in the tropics has increased the salinity. Low salinity in coastal areas is due to the influx of freshwater from rivers.*

Rainwater tends to be slightly acidic due to the presence of dissolved carbon dioxide and sulfur dioxide. When it falls on rocks, this water breaks them down, releasing mineral salts that are washed into rivers and, eventually, the oceans. Over billions of years, these salts have built up in the sea, making it salty.

The average salt content of seawater is about 3.5 percent, or 35 parts per thousand, which adds up to about 55 million billion tons (50 million billion tonnes) of salt in total. Most of the salt is the same as the salt we use for cooking; sodium and chloride make up about 85 percent of all the ions found in seawater, with magnesium and sulfate making up another 10 percent. The concentrations of the different ions are affected by biological activity. For example, calcium and carbonate ions are taken up by coral polyps to build their skeletons and remain behind in the form of coral reefs.

Local conditions can have a significant effect on seawater's salinity. In the tropics, higher temperatures lead to more evaporation, leaving the water saltier. However, high volumes of rainfall can also dilute the surface water in equatorial regions. The picture is more complex at the poles: when sea ice freezes, it increases the salinity of the surrounding water as the salt is left behind; when it melts, it dilutes the seawater, reducing its salinity. There is also little evaporation at the poles.

Warm ocean regions with lower circulation, such as the Mediterranean Sea, tend to have higher salinities. The most saline area of open sea is the Red Sea, which experiences high rates of evaporation and low precipitation, as well as confined circulation and little in the way of river run-off. The seas off Southeast Asia tend to have lower salinities due to high levels of precipitation and high volumes of water flowing out from rivers.

The saltier water is, the denser it becomes. Salinity-influenced differences in water density help to drive the ocean conveyor belt, which moves seawater around the planet (see page 110). The average density at the ocean's surface is 1.28 pounds per pint (1.025 kilograms per liter).

The freezing point of seawater decreases with increasing salinity. Typically, seawater freezes at about 28°F (-2°C).

Below *Arctic sea ice. When sea ice forms, salt is expelled, increasing the salinity of the surrounding water, which then decreases its freezing point.*

// Sea level

The height of the world's oceans in relation to the land is not static, changing over both short and geological time scales.

Because the sea is in constant motion, its height compared to the land is constantly changing. Numerous factors affect sea level, including tides, winds, ocean currents, atmospheric pressure, local gravitational differences, and water temperature and salinity, making its precise determination difficult. Hence, we mostly talk about the mean sea level for a given location.

Mean sea level is generally determined by selecting a location and taking hourly measurements of the sea level over a 19-year period then taking the average of the measurements. This allows for the fluctuations caused by waves and tides to be taken into account, as well as the longer-term changes caused by the effects of the Metonic cycle, whereby the Moon's phases match up with the days of the solar year over a 19-year period. Today, satellite altimeters are used to make precise measurements of sea level.

Because sea level is measured in relation to the adjacent land, changes in the height of the land itself can also have an influence. The weight of an ice sheet, for example, can force the land to move downwards. When that ice sheet melts, the land slowly bounces back, in some cases rising more than 330 feet (100 meters), a phenomenon known as isostatic rebound.

Sea levels have varied greatly over geological time scales. During periods when the climate was much colder, large amounts of water were frozen into glaciers and vast ice sheets, significantly lowering the sea level. For example, about 18,000 years ago, when the most recent glacial period was at its peak, mean sea level was about 330 feet (100 meters) lower than it is today. At that time, the continental shelves between North America and Asia were probably exposed and it is believed that they formed a land bridge between the two continents, which could have been used by humans to migrate from Siberia to what is now Alaska, becoming the first people in North America.

Since then, sea levels have been rising as the climate has warmed—first naturally and more recently due to anthropogenic climate change. For at least the past century, sea level has been rising by about 0.07 inches (1.8 millimeters) per year. Melting of glaciers and ice sheets, the expansion of water as it warms and the extraction of groundwater for agriculture and other uses have all played a part in that rise.

Left *A tide marker stands exposed on a beach at low tide.*

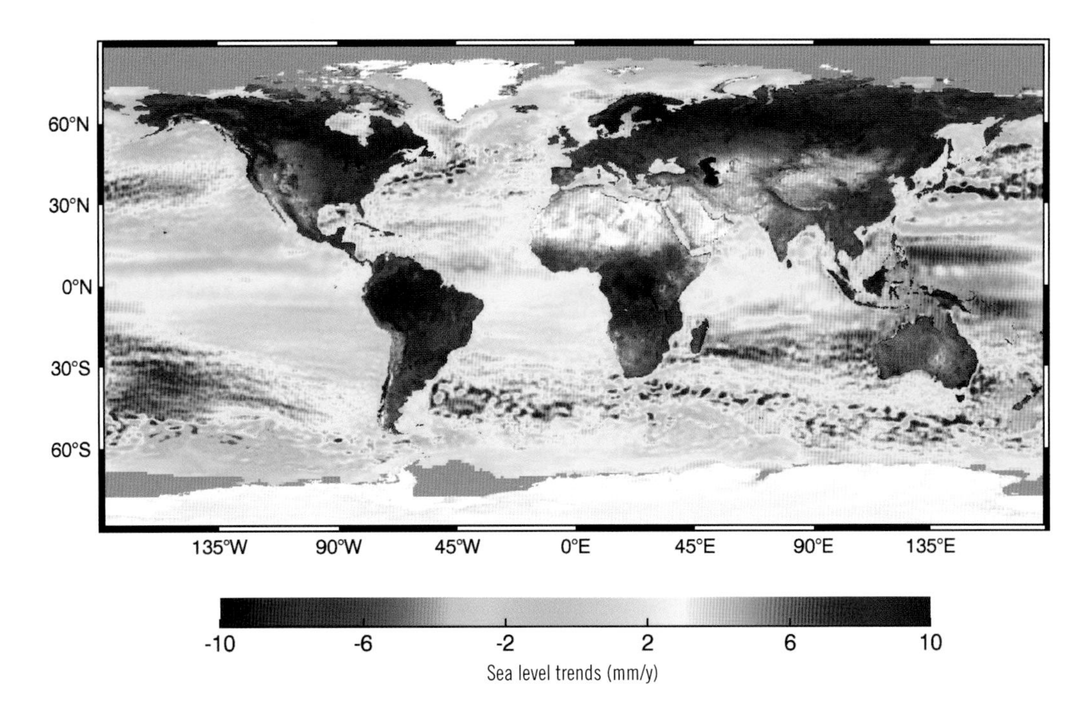

Left *Trends in global sea level between January 1993 and January 2016. The size and direction of sea-level change varies from place to place due to local factors such as subsidence, post-glacial rebound, erosion, and regional ocean currents. Thermal expansion of water due to heating can also have a local impact.*

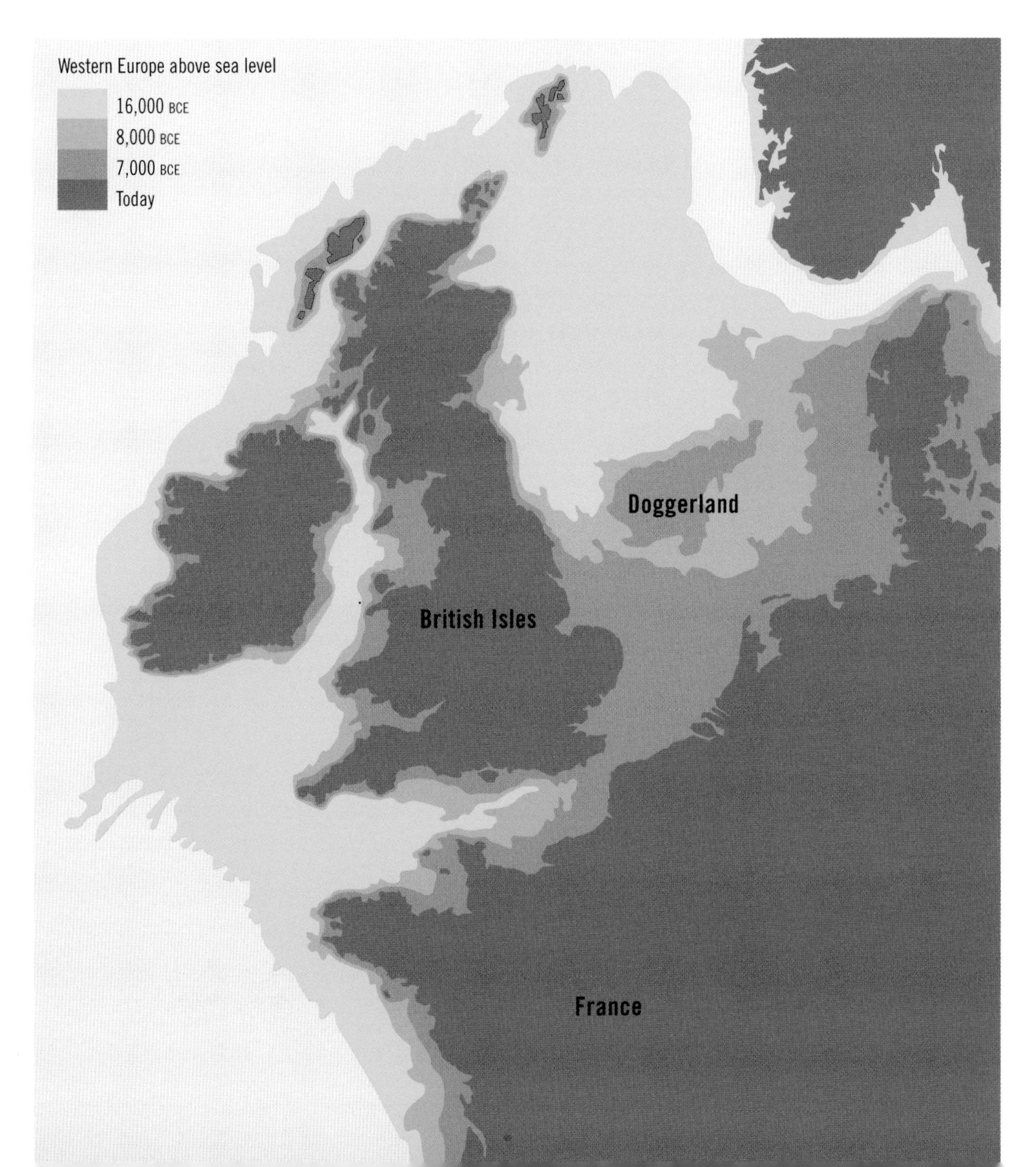

Left *During the last glacial period, so much water was locked up in ice sheets and glaciers that sea levels were about 330 feet (100 meters) lower than they are now, which meant that large areas of the continental shelf were exposed. Among these was a region now known as Doggerland, which connected the UK and Ireland to continental Europe. It was flooded by rising sea levels in around 6500–6200 BCE.*

// Tides

Each day, the level of the sea fluctuates along the world's coastlines. These predictable rises and falls, known as tides, are caused by the gravitational attraction of the Moon and Sun on the water in the oceans.

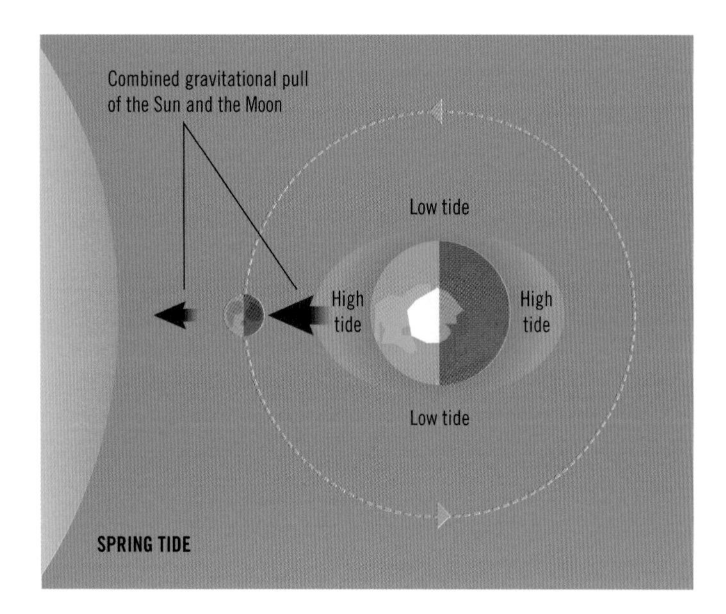

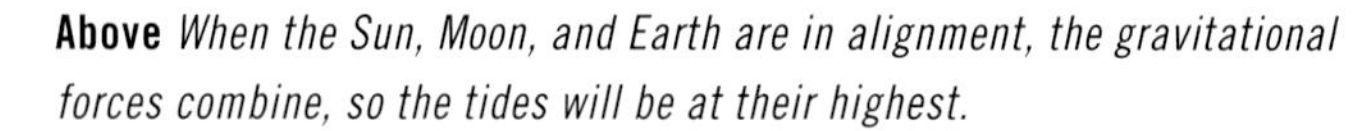

Above *When the Sun, Moon, and Earth are in alignment, the gravitational forces combine, so the tides will be at their highest.*

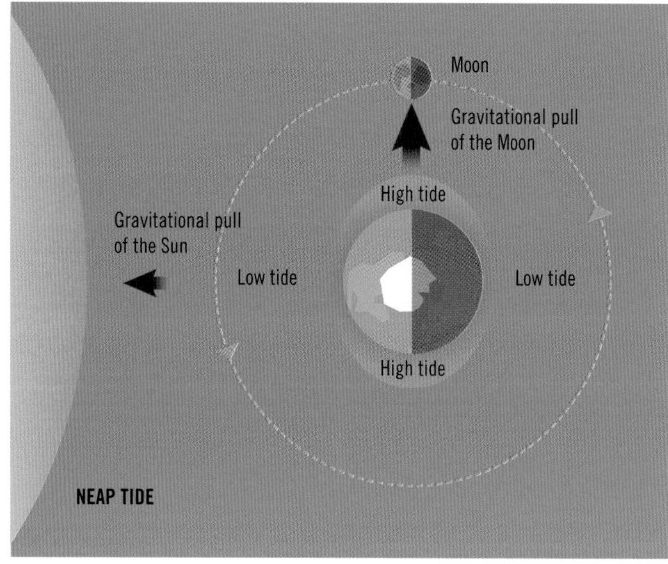

Above *When the Earth, Moon, and Sun form a right angle, their gravitational pulls work against each other, so the tides are at their lowest.*

Although both the Moon and the Sun have an impact on the tides, the Moon is much closer to the Earth, so its effect is about twice that of the Sun. The Moon's gravity pulls the water upwards on the side of the Earth that faces it. On the far side of the Earth, away from the gravitational effect of the Moon, the water tends to "fly out" from the Earth due to inertial forces. In the open ocean, this results in two bulges of water, one that is adjacent to the Moon and another on the opposite side of the Earth. Between these two bulges are troughs of lower water. As the Earth rotates beneath these bulges, areas along the coast experience a rise (flow) and fall (ebb) of the water level—the tides. Each high tide is separated from the next by about 12 hours and 25 minutes.

NEAP AND SPRING TIDES

The tidal range at any given place varies over time as the gravitational interactions of the Sun, Moon, and Earth change, rising to a maximum during what's known as a spring tide and dropping to a minimum during a neap tide. Spring tides occur twice a month, after every full and new Moon, when the Sun, Moon, and Earth are in alignment and the gravitational force is at its strongest. They alternate with neap tides, which take place when the Earth, Moon, and Sun form a right angle, so the gravitational forces are working against each other—this is when the Moon is in the first and third quarter. The largest tidal range of the year takes place when a spring tide coincides with an equinox.

Effect of geography

Geography affects the tidal range experienced by a coastline as well as the pattern of the tides themselves. For example, the timings of high and low tides can vary by several hours in areas that are relatively close to each other; and in some parts of the world, local effects can mean that there is only one tide a day or even none.

Seafloor features, in particular, can have a significant impact on tidal ranges. In areas adjacent to shallow continental shelves, the height of the tide is usually magnified, while mid-oceanic islands that rise steeply from the sea floor have smaller tidal ranges. Similarly, in narrow-mouthed basins, the tides often rise higher than they do in wide bays. In estuaries, the shape of the river bed may alter the tidal pattern, with long flood times and

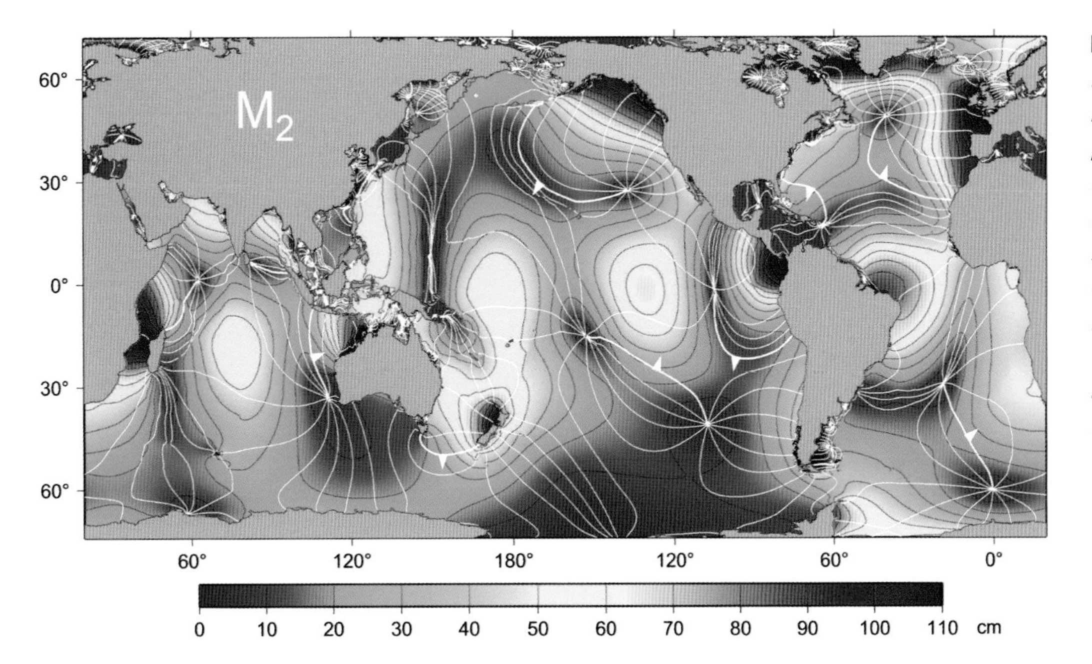

Left *This map shows the tidal range attributable to the so-called M2 tidal constituent—that is, the part of the tide that is related to the direct gravitational effect of the Moon. The regions marked in red show the greatest extremes (the highest highs and lowest lows), while the places where the white lines converge are so-called amphidromic points, where the tidal range is zero.*

short ebb times or vice versa. In such cases, the low tide will not necessarily be midway between two high tides.

The magnitude of the effect of the Moon's gravitational pull depends on the volume of water on which it is acting. Hence tidal ranges tend to be greatest around the Pacific Ocean, because its basin is the largest and holds the greatest volume of water. In contrast, smaller water bodies that are relatively enclosed may experience very small tidal ranges. The Mediterranean, Baltic, and Caribbean seas, for example, experience some of the smallest tidal ranges. A location where the tidal range is zero is known as an amphidromic point or tidal node.

The greatest tidal range occurs in the Bay of Fundy, Canada, where the difference between the high and low tide is 53.5 feet (16.3 meters). In the UK, the Severn Estuary between England and Wales regularly experiences tidal ranges of up to 50 feet (15 meters).

Below *An unusually high tide submerges Saint Mark's Square in Venice, Italy. The so-called* acqua alta *most often takes place between fall and spring, when the astronomical tide is enhanced by the seasonal prevailing winds.*

// Ocean zones

Oceanographers classify the ocean into a series of zones, both horizontally and vertically.

Horizontally, the ocean is divided up based on the distance from the shoreline. Closest to the shore is the intertidal or littoral zone, which incorporates the area between the high- and low-tide marks. This is an area of constant change, affected by the rising and falling tides, wave action, and longshore currents. Beyond the intertidal zone lies the neritic zone, which extends out to the seaward edge of the continental shelf. The remainder is known as the oceanic zone, which contains about 65 percent of the ocean's completely open water. Together, the neritic and oceanic zones are termed the pelagic zone. The sea floor itself, along with any sediments lying on it, is called the benthic zone.

Vertically, the ocean is divided into five zones according to water depth (see diagram). Three main factors change with increasing depth: the amount of available light, the temperature, and the water pressure.

As sunlight percolates below the surface, it is attenuated as it is absorbed by water molecules and scattered by molecules and particulates suspended in the water. Different wavelengths of light are attenuated at different rates. Both short (ultraviolet) and longer wavelengths (red to infrared) are rapidly absorbed, leaving only light at the blue-green end of the spectrum to penetrate to any depth.

The layer of water through which light can penetrate is known as the photic zone; below this lies the aphotic zone. The uppermost part of the photic zone, where the light intensity is sufficient for plants to conduct photosynthesis—usually the top 330 feet (100 meters) or so—is known as the euphotic zone; the lower portion of the coean, where there isn't enough available for photosynthesis, is called the dysphotic zone. In the dysphotic zone and below, many organisms make their own light—typically for luring prey or for navigation—using a process known as bioluminescence. This light is typically blue-green in color.

Nearly all primary production in the ocean occurs in the euphotic zone. Organisms that live in the aphotic zone either migrate upwards to feed, prey on other creatures around them or rely on material sinking from above.

Water temperature also decreases with depth. The upper layers, which are warmed by the Sun, can reach temperatures of more than 95°F (35°C), particularly when they are still and don't mix with cooler water below.

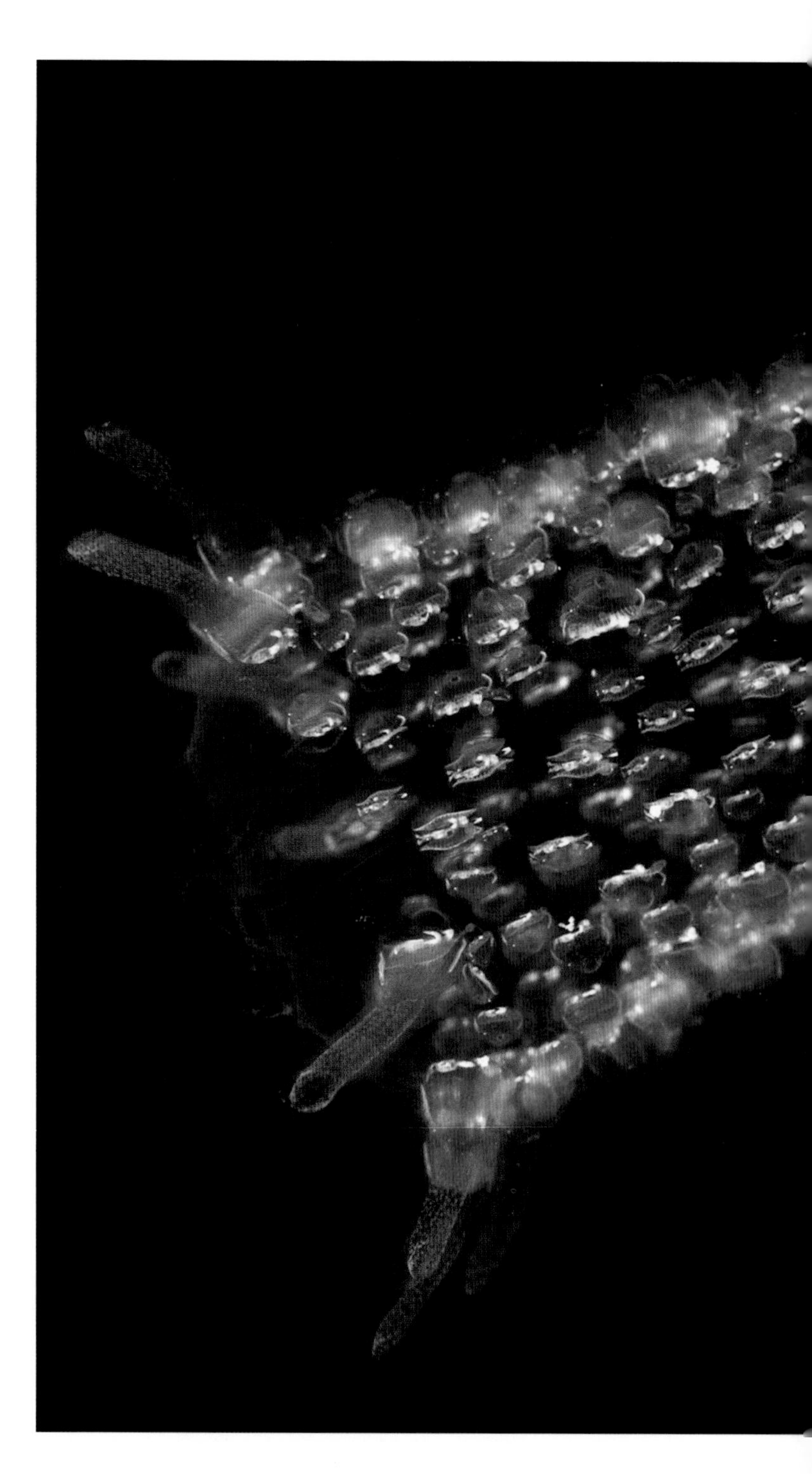

Regions of the water column that undergo dramatic changes in temperature with depth are known as thermoclines. The depth and strength of a thermocline will vary over time, depending on factors such as the amount of mixing and the surface temperature, and with latitude; the tropical thermocline is typically stronger and deeper than the thermocline at higher latitudes. Because they receive relatively little solar energy and so are as cold as deep-ocean water, polar waters tend to lack a thermocline.

THE OCEAN ZONES

Epipelagic Zone (0–650 feet/0–200 meters). Also known as the sunlight zone, the epipelagic zone is both the warmest layer and the one with the greatest temperature variation. It is in this layer that most biological activity takes place. Winds at the surface tend to keep the water here well mixed, allowing heat from the Sun to penetrate to deeper levels. Temperatures: 28°F–95°F (-2°C–35°C)+

Mesopelagic Zone (650–3,300 feet/200–1,000 meters). Sometimes referred to as the twilight or midwater zone. Light still penetrates this zone, but it is extremely faint and not sufficient for plants to photosynthesize; bioluminescent creatures begin to appear. In tropical and temperate regions, a permanent thermocline is found within this layer. Temperatures: 39°F–68°F (4°C–20°C)

Bathypelagic Zone (3,300–13,000 feet/1,000–4,000 meters). Sometimes referred to as the midnight zone or the dark zone. The only visible light present within this zone is produced by bioluminescent organisms. A little more than half of the ocean lies within this zone. Temperature: 39°F (4°C)

Abyssopelagic Zone (13,000–20,000 feet/4,000–6,000 meters). Also known as the abyssal zone or simply the abyss. Three quarters of the ocean floor is found within this zone, but there are very few sea creatures. Temperatures: 28–39°F (2°C–4°C)

Hadalpelagic Zone (20,000 feet/6,000 meters and below). Also known as the hadopelagic zone. Waters in this zone are mostly found in trenches and canyons. Very few sea creatures live at these depths. Temperatures: 34–39°F (1°C–4°C)

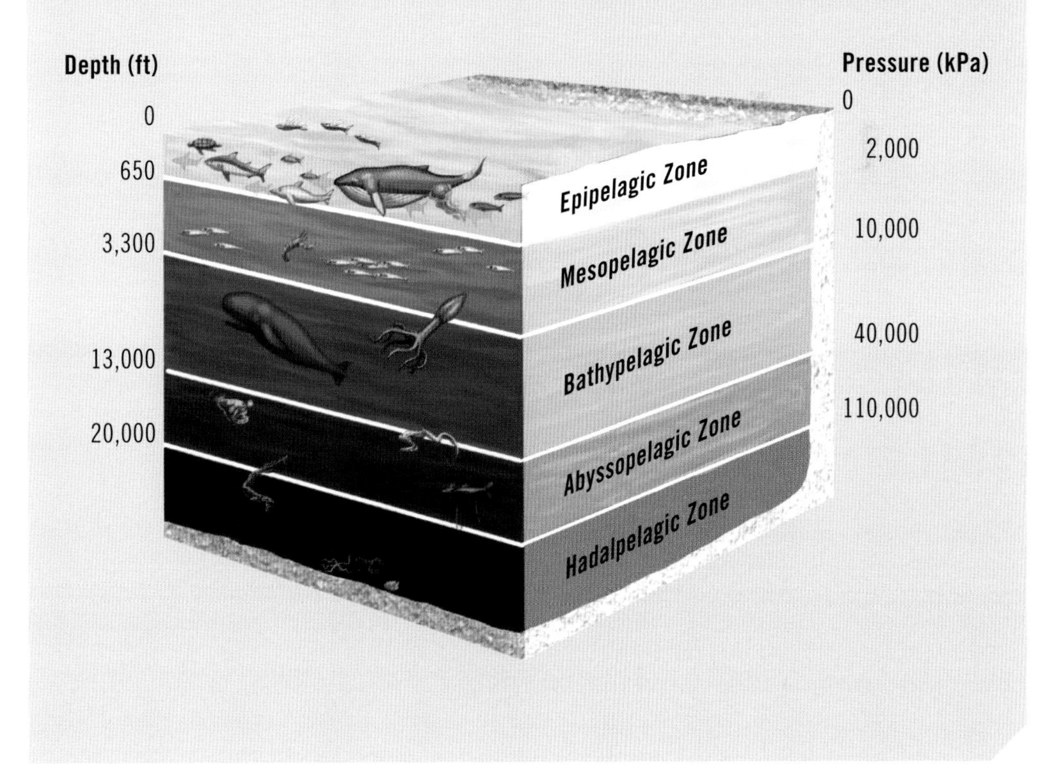

Above *A bioluminescent pyrosome, a type of colonial tunicate, near Atauro Island, East Timor. Unlike most luminescent plankton, pyrosomes generate brilliant, sustained bioluminescence and they are among the few marine organisms reported to luminesce in response to light.*

// Continental shelves

Around the edges of the continents are broad, flat areas submerged by the sea—the continental shelves.

Extending out from the land for an average of about 45 miles (70 kilometers), continental shelves typically slope slightly seaward at an angle of about 0.1°. The topography of the shelf is usually quite gentle, with small hills and ridges alternating with shallow depressions and valley-like troughs, a landscape known as ridge and swale. In some places, however, they are dissected by deep submarine canyons. These are often located near river mouths, forming as water flowing out into the ocean erodes the underlying material. The Congo Canyon, at the mouth of the Congo River, is 500 miles (800 kilometers) long and 4,000 feet (1,200 meters) deep. The geology of the continental shelf tends to be similar to that of the adjacent exposed portion of the continent.

The broad submarine terrace of the continental shelf eventually drops away at a point called the shelf break, which is usually located at a depth of about 460 feet (140 meters). The steep section found on the seaward side of the shelf break is known as the continental slope. At its base is a sediment-filled region called the continental rise that gently slopes down to the abyssal plain (see page 96). The continental rise may extend as far as (300 miles/500 kilometers) from the continental slope. The continental shelf, slope, and rise are collectively known as the continental margin.

Continental shelves are usually covered by a layer of sediment that has been worn away from the exposed part of the continent by erosion. As it is washed off the shelf,

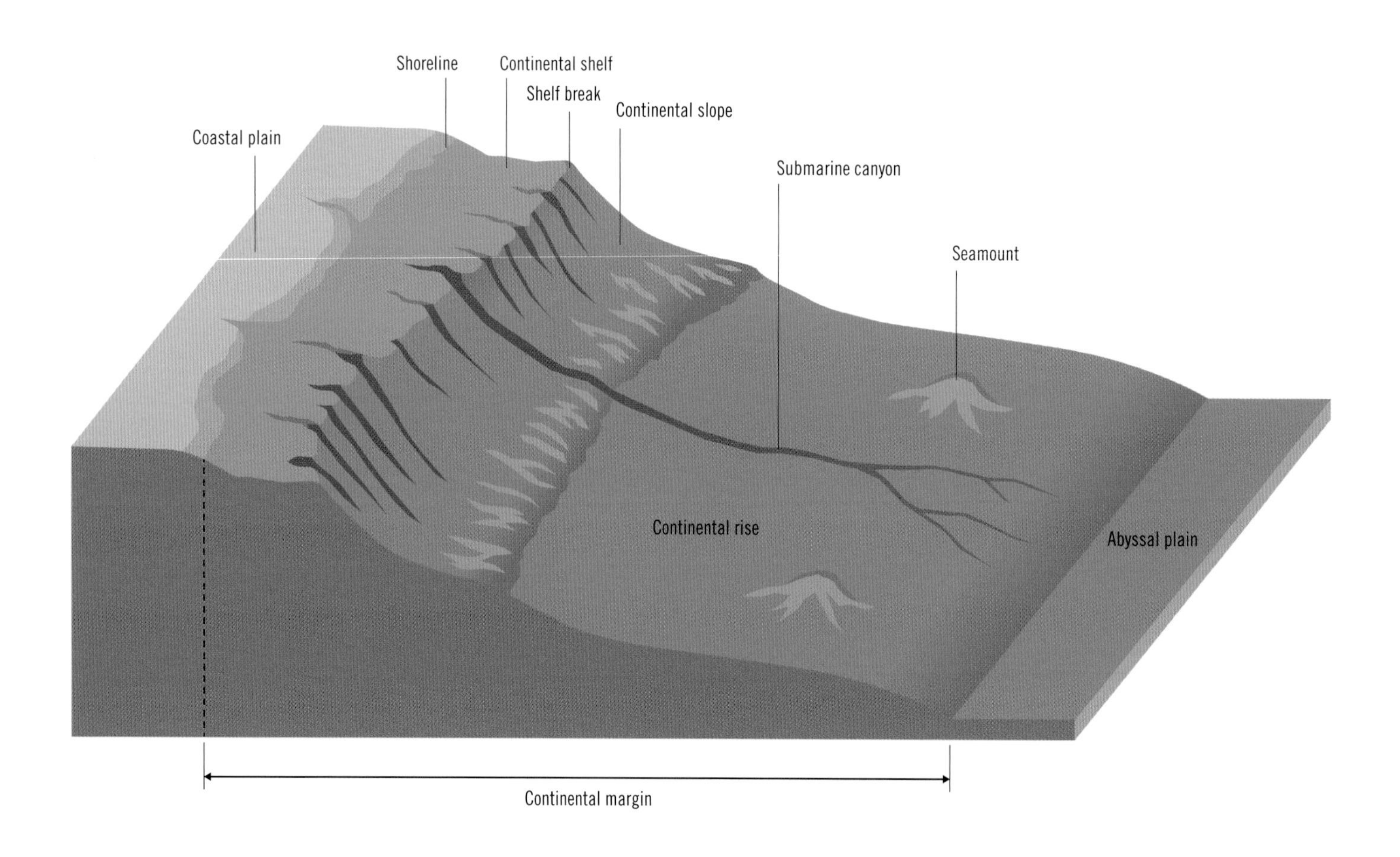

Below *The relatively flat continental shelf drops away at the shelf break into a steep section known as the continental slope. At its base is a gentler-sloped, sediment-filled region called the continental rise, which eventually gives way to the flat abyssal plain. Together, the continental shelf, slope, and rise are known as the continental margin.*

this sediment accumulates in the continental rise, before eventually slumping down onto the abyssal plain.

Continental shelves vary considerably in width. In regions where an oceanic plate is being pushed under a section of continental crust, such as off the coasts of Chile and Sumatra, there may be virtually no shelf at all. Elsewhere, the shelf may extend for hundreds of miles. The world's largest continental shelf is on the coast of Siberia, reaching more than 750 miles (1,200 kilometers) into the Arctic Ocean. Together, continental shelves comprise about 8 percent of the total area of the world's oceans.

During glacial periods, when sea levels drop, continental shelves may be exposed, potentially creating land bridges that allow animals—including humans—to migrate between landmasses that are normally separated by sea.

Above *World map showing the depths of the oceans. The continental shelves can be seen as pale areas fringing the different landmasses.*

The ocean water above the continental shelf has an average depth of about 200 feet (60 meters). These areas are known as shelf seas; examples include the South China Sea, the North Sea and the Persian Gulf.

Continental shelves tend to be highly productive. Sunlight penetrates the relatively shallow, warm water, allowing algae and other plants to flourish, and ocean currents and river runoff provide nutrients.

The continental shelf has great economic significance for countries that border the sea. Most of the off-shore mining of oil, gas and other minerals takes place on the continental shelf, as does much of the commercial fishing.

// Abyssal plains

In the deep ocean, between the edges of the continents and the great underwater mountain ranges of the plate boundaries, lie vast flat areas known as abyssal plains, among the flattest, smoothest, least-explored regions on Earth.

As new sea floor is created at mid-ocean ridges, it is slowly blanketed by fine-grained sediment, mainly clay and silt. It is this sediment that make the abyssal plains so flat (the depth across an abyssal plain only varies by about 4-40 inches/10–100 centimeters per half a mile [just under a kilometer] of horizontal distance), the originally uneven surface of the oceanic crust being filled in by the falling sediment. In remote areas, sediment is deposited at a rate of about 1 inch (2.5 centimeters) per 1,000 years.

Much of the sediment has flowed off the continental shelf and accumulated on the continental slope. When this sediment becomes unstable, it slumps down and flows out in what are known as turbidity currents. These are often channeled along submarine canyons and out over the adjacent abyssal plain. The remainder of the sediment is made up mostly of dust that has been blown out to sea from the land and what are known as pelagic sediments—the remains of small marine plants and animals that have sunk down from the ocean's upper layers. The sediment cover can be more than a mile deep. In some areas, the plains are dotted with potato-sized nodules containing metals such as manganese, iron, nickel, cobalt, and copper.

Usually found at depths of between 10,000 and 20,000 feet (3,000–6,000 meters), and generally adjacent to a continent, most abyssal plains are stretched out along the continental margin. They may be hundreds of miles wide and thousands of miles long. The Sohm Plain in the North Atlantic has an area of about 350,000 square miles (900,000 square kilometers).

Taken together, abyssal plains may account for almost a third of the Earth's surface—about as much as all of the dry land combined. They are most common in the Atlantic Ocean, where they also reach their largest extent. They are rarest in the Pacific, where deep trenches near the continental margins trap most of the sediment before it reaches the open ocean.

It was once assumed that the abyssal plains were largely devoid of life, but recent research has revealed that they teem with a wide variety of microbes, including up to 2,000 species of bacteria and 250 species of protozoa. The sediment may also contain amoeba-like single-celled organisms called foraminifera. Invertebrate biodiversity can be high, too, particularly around hydrothermal vents and cold seeps (see page 100).

Organisms that live on the abyssal plains must contend with severe food limitations, relying mostly on the tiny particles of organic material that rain down from the water near the surface, the occasional large carcass that sinks down and organic material that flows down off the continental margins.

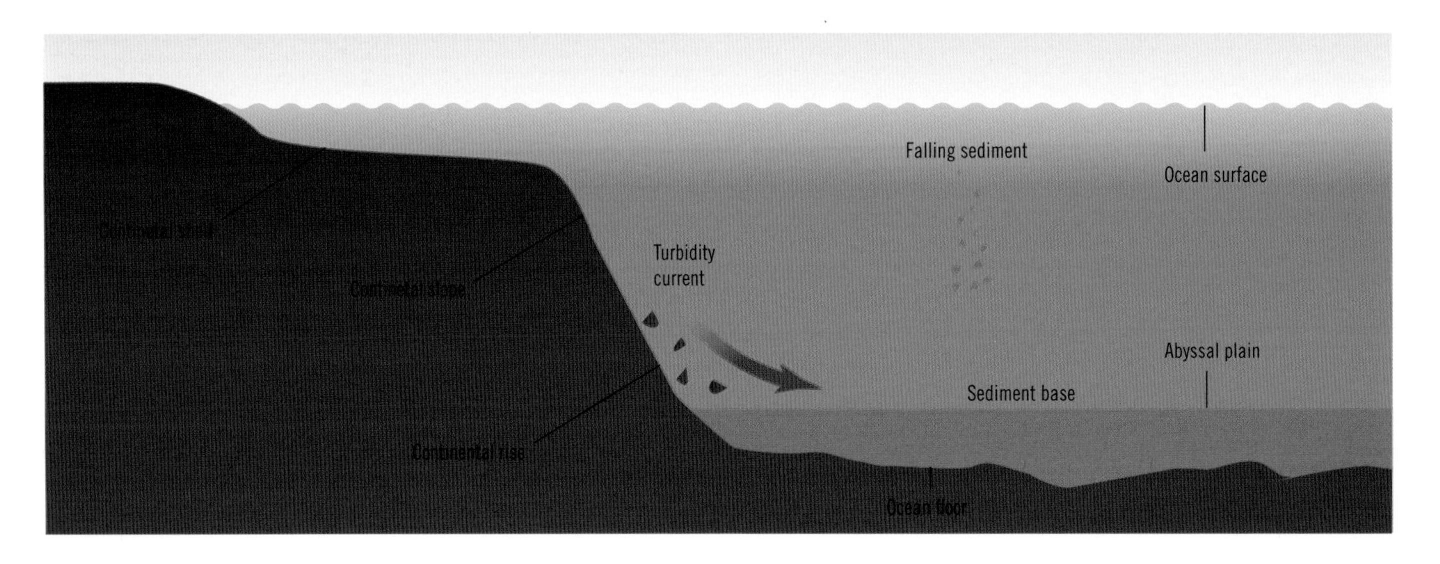

Below *Sediment falling from the ocean's upper levels and sweeping down in turbidity currents from the continental slope fills in hollows in the ocean floor, making the abyssal plain extremely flat.*

A bathysaurus sits on an abyssal plain surrounded by brittle stars.

// Seamounts

Just as the land is populated by vast, towering mountain ranges, so too are the world's oceans.

Over millennia, volcanic activity has led to the formation of underwater mountains that can rise thousands of feet above the sea floor. Those that are more than 3,300 feet (1,000 meters) tall are known as seamounts (smaller submarine volcanoes are called sea knolls).

Like terrestrial volcanoes, seamounts are generally conical in shape; however, many have flat tops, in which case they are known as guyots. Guyots are the remnants of undersea volcanoes that once reached above the sea surface. Wave action slowly eroded away their summits, creating a flat surface, before subsidence eventually caused them to sink back below the surface.

Seamounts are present in all of the world's major ocean basins, but are most common in the Pacific; 60 percent of known examples are found there. They are abundant—more than 14,000 have been identified, covering about 5 percent of the ocean floor—but little explored.

Like volcanoes on land, seamounts typically occur either near the boundaries of the Earth's tectonic plates or in mid-plate areas over magma hotspots. At mid-ocean ridges, where the continental plates are moving away from each other, magma rises to fill the gap. If this happens rapidly enough, an undersea volcano forms. Similarly, near subduction zones, where the plates are colliding, the oceanic crust is forced down into the Earth's scorching interior. When it melts, it forms magma that rises back to the surface and erupts via volcanoes.

Below *Seamounts in the northern part of the Mariana Arc, a crescent-shaped chain of numerous seamounts and a few islands that extends from Guam to Japan. The seamounts, which are all active volcanoes, are fed by melting of the Pacific Plate, which is subducting beneath the Philippine Plate at the nearby Mariana Trench. The image is three times vertically exaggerated.*

Above *A large primnoid coral covered in orange brittle stars on Dickins Seamount in the Gulf of Alaska. Because their upper reaches often lie close to the sea surface, seamounts frequently support diverse communities of sea life.*

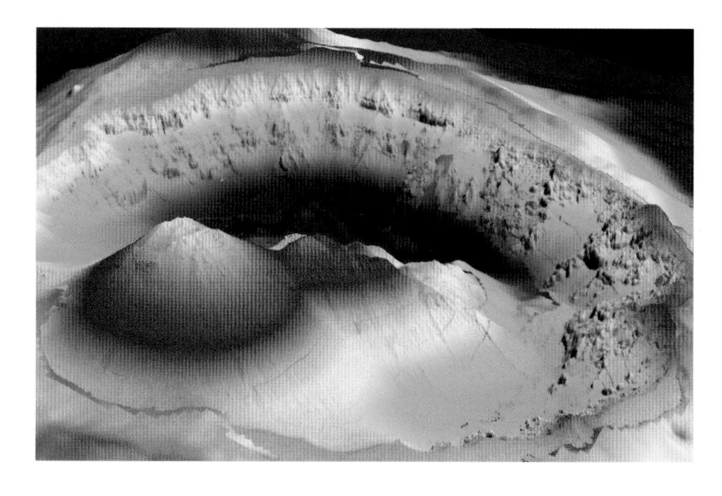

Above *The Brothers Seamount is an active volcano located in the Pacific Ocean off New Zealand. The caldera is about 1.8 miles (three kilometers) wide and contains a smaller active cone that rises about 1,150 feet (350 meters) above the caldera floor. The walls of the caldera and the volcanic cone host numerous hydrothermal vents.*

Seamounts are often found in linear clusters or elongated groups, formed as lava is extruded from a linear rift in the crust. The clusters may contain as many as 100 seamounts strung out over thousands of kilometers.

Some seamounts can be enormous. Great Meteor Tablemount, in the northeast Atlantic, is up to 70 miles (110 kilometers) across at the base and rises more than 2.5 miles (4 kilometers) above the surrounding sea floor. As they sometimes reach close to the ocean surface, they can present a danger to shipping. However, the greatest risk comes from the possibility of underwater landslides caused by flank collapses, which can potentiall y generate massive tsunamis.

Underwater islands of life

Seamounts support some the world's most productive marine ecosystems. Because they provide a surface on which animals and plants can settle, reach up towards

Above *Part of the Lord Howe Seamount Chain, a series of submerged mountains that runs for 621 miles (1,000 kilometers) from the Coral Sea to Lord Howe Island, Gifford Guyot is 6,562 feet (2,000 meters) high; its summit is about 820 feet (250 meters) below the surface.*

the ocean's surface and create physical obstacles that force deep-water ocean currents bearing nutrient-rich waters upwards, seamounts are fertile habitats that are home to diverse communities of marine life. Many are productive fishing grounds, collectively providing more than 80 commercial species of fish and shellfish worldwide. However, there are concerns that fishing is having a negative impact on seamount ecosystems, particularly when bottom trawling takes place.

There is evidence that migratory sea life such as whales and whale sharks use seamounts as navigational aids.

THE WORLD'S TALLEST MOUNTAIN

No, it is not Mount Everest. Hawaii's Mauna Kea, a dormant volcano, stands more than 33,000 feet (10,000 meters) tall, rising 13,802 feet (4,207 meters) above the sea surface and dropping a further 20,000 or so feet (6,000 or so meters) down to the sea floor.

// Hydrothermal vents

Like submarine geysers, hydrothermal vents spew scalding water into the deep ocean, in the process supporting communities of bizarre sea creatures and creating a rich bounty of precious metals.

First discovered in 1977, hydrothermal vents typically form along mid-ocean ridges, where two tectonic plates are spreading apart and new crust is being formed as magma rises to the surface; they are also sometimes found around submarine volcanoes. Seawater circulates within the crust through cracks and porous rocks, and is heated to extreme temperatures by the molten magma before exiting through vents in the surface. While the surrounding water may only be 36°F (2°C), the water emerging from the vents may be as hot as 867°F (464°C); however, it doesn't boil because it is under extreme pressure.

As it percolates through the crust, the superheated water picks up dissolved minerals. Chemical reactions driven by the heat make the water more acidic, causing it to leach metals such as iron, zinc, copper, lead, and cobalt from the surrounding rocks. When this water leaves the vent and mixes with the near-freezing surrounding water, a new set of chemical reactions rapidly take place and the minerals come out of solution and solidify, creating complex structures around the vent.

Some vents release water that is high in sulfide minerals. These solidify into fine-grained black particles, making it look as though the vents are chimneys spewing out smoke—hence their name: "black smokers." There are also "white smokers," which are cooler and release particles of barium,

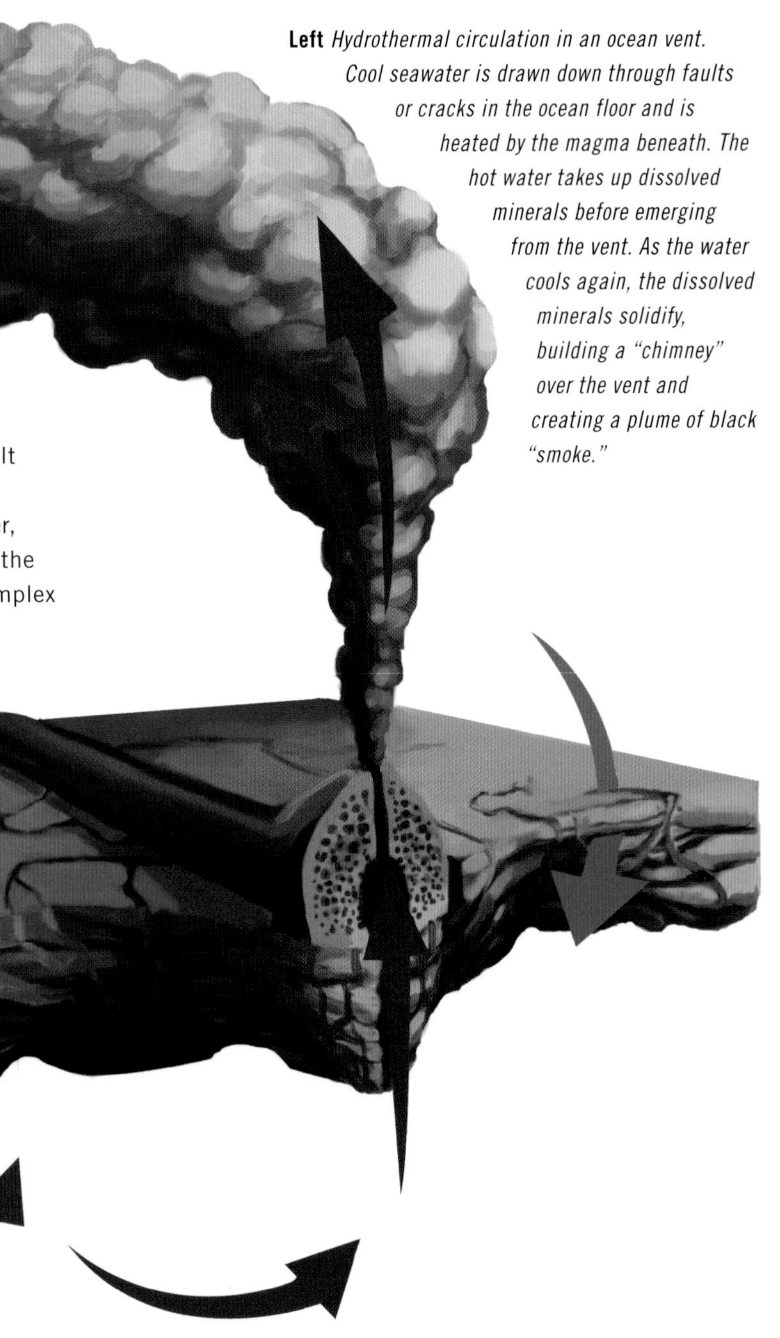

Left *Hydrothermal circulation in an ocean vent. Cool seawater is drawn down through faults or cracks in the ocean floor and is heated by the magma beneath. The hot water takes up dissolved minerals before emerging from the vent. As the water cools again, the dissolved minerals solidify, building a "chimney" over the vent and creating a plume of black "smoke."*

Above *A black smoker located over a mid-ocean-ridge hydrothermal vent in the Atlantic Ocean. The "smoke" emerging from the chimney-like structure is mostly made up of fine-grained sulfide minerals, formed as the hot hydrothermal fluid mixes with the surrounding seawater, which is close to freezing temperature.*

calcium, and silicon. Black smokers are typically found at depths of 8,000–10,000 feet (2,500–3,000 meters), often in groups spread over hundreds of square meters. Their chimneys can grow at rates of up to 12 inches (30 centimeters) a day and can reach heights of 200 feet (60 meters).

Hydrothermal vents often support diverse biological communities; the density of organisms around them can be 100,000 times greater than that of the surrounding sea floor. Unlike in most ecosystems, the primary producers in these communities rely not on photosynthesis but on chemosynthesis—in this case, bacteria use chemicals such as hydrogen sulfide as an energy source. These bacteria then support a wide range of other organisms, including tubeworms, crustaceans, snails, mussels, and even fish. Some of these organisms graze on mats of bacteria, some have formed symbiotic relationships with the bacteria, hosting them within their tissues, and others simply prey on the other organisms that live around the vents.

COLD SEEPS

Tectonic plate boundaries also often host vents known as cold seeps, where much cooler water that is rich in hydrocarbons flows from the sea floor. Cold seeps are not particularly cold, indeed the water they release is often slightly warmer than their surroundings. They have been found in all of the major oceans at a wide range of depths—in Chile, they occur in the intertidal zone—but they are mostly in deeper water. In contrast to hydrothermal vents, which are typically volatile and short-lived, seeps tend to be more stable, emitting fluid at a slow and dependable rate.

Like hydrothermal vents, cold seeps support biological communities. In this case, the chemosynthetic bacteria that form the base of the ecosystem use methane and hydrogen sulfide as an energy source. The worms that live around cold-seep environments are among the world's longest-living invertebrates, reaching ages of up to 250 years.

Above *Chemosynthetic mussels and squat lobsters near a hydrothermal vent on the NW Eifuku volcano in Japan's Volcano Island chain. Although small, the volcano is extremely active, with numerous white smokers that emit hydrothermal fluid at a temperature of about 221°F (105°C) that includes liquid carbon dioxide bubbles, one of only two places where this has been observed.*

// Seafloor spreading

Where continental plates are pulling away from each other, new crust is constantly being formed, a process known as seafloor spreading.

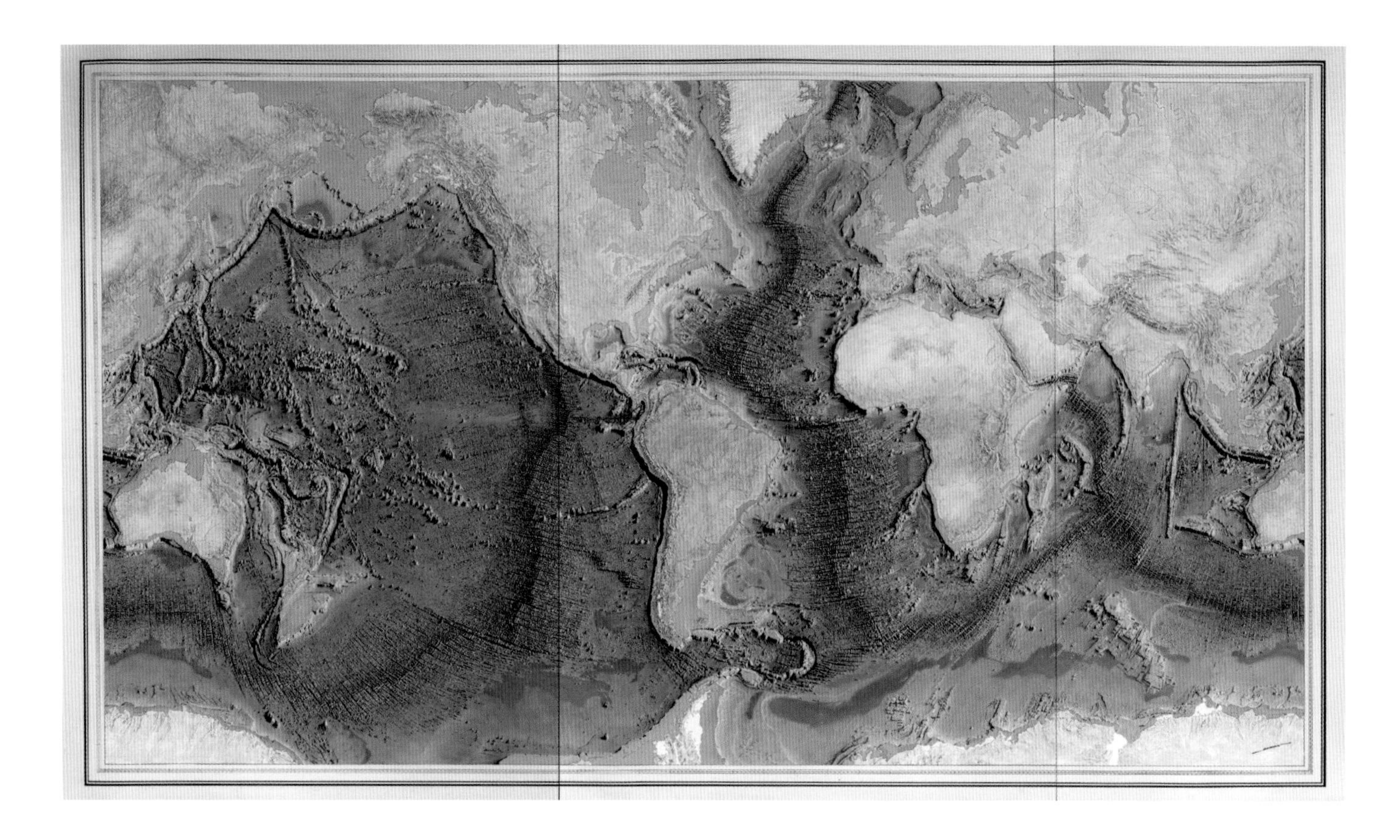

Above *This hand-painted map by Austrian painter and cartographer Heinrich C. Berann from 1977 clearly shows the mid-ocean ridges that run through the Earth's major ocean basins.*

Seafloor spreading initially takes place at the boundaries of tectonic plates that are moving away from each other, forming a rift. Mantle rock, heated by the Earth's core, rises up below the rift. As it rises, the reduction of pressure lowers the rock's melting point, allowing it to liquify and become magma, a process known as decompression melting. This magma rises to the surface and fills the newly formed gap between the plates before solidifying to form new oceanic crust.

The continents that border the Atlantic Ocean are thought to be moving away from the Mid-Atlantic Ridge at a rate of 0.4 to 0.8 inches (1–2 centimeters) per year. Hence, the Atlantic Ocean basin is widening by 0.8 to 1.6 inches (2–4 centimeters) per year.

The newest, thinnest crust is found near the centre of a mid-ocean ridge. As the newly created sea floor cools, it becomes denser and so sinks slightly, so older oceanic basins are deeper than younger ones. As you move away from a mid-ocean ridge, the age, density and thickness of the oceanic crust all increase. Oceanic crust is eventually consumed through subduction, so it is rarely more than 200 million years old.

Sometimes, seafloor spreading leads to the creation of new geographic features. The divergence of the African and Arabian plates, for example, created the Red Sea and will eventually join it to the Mediterranean.

The rate of creation of new sea floor is not the same at all of the mid-ocean ridges. The Mid-Atlantic Ridge is spreading at a rate of 0.8 to 2 inches (2–5 centimeters) a year, whereas the East Pacific Rise, a mid-ocean ridge that runs through the eastern Pacific Ocean, is spreading by about 2.4 to 6.3 inches (6–16 centimeters) a year. The slower-spreading ridges typically feature tall, narrow underwater cliffs and mountains, while more rapidly spreading ridges have much gentler slopes.

This movement of the sea floor can lead to the formation of island chains. The Hawaiian Islands, for example, were formed as the sea floor slowly moved over a hotspot or mantle plume.

The world's mid-ocean ridges are all connected. Together, they wind for nearly 50,000 miles (80,000 kilometers) through all the world's oceans, forming a continuous 40,000-mile (65,000-kilometer) mountain range, the world's longest.

Below *Evidence for seafloor spreading can be seen in magnetic "stripes" found in oceanic crust. Before rising magma solidifies, iron-based minerals within it are imprinted with the polarity of the Earth's magnetic field at the time. This is then literally set in stone when it hardens into oceanic crust, creating a band that runs parallel to the mid-ocean ridge. When the magnetic field's polarity reverses, a new stripe is created. As the sea floor spreads, the matching stripes move away from the ridge.*

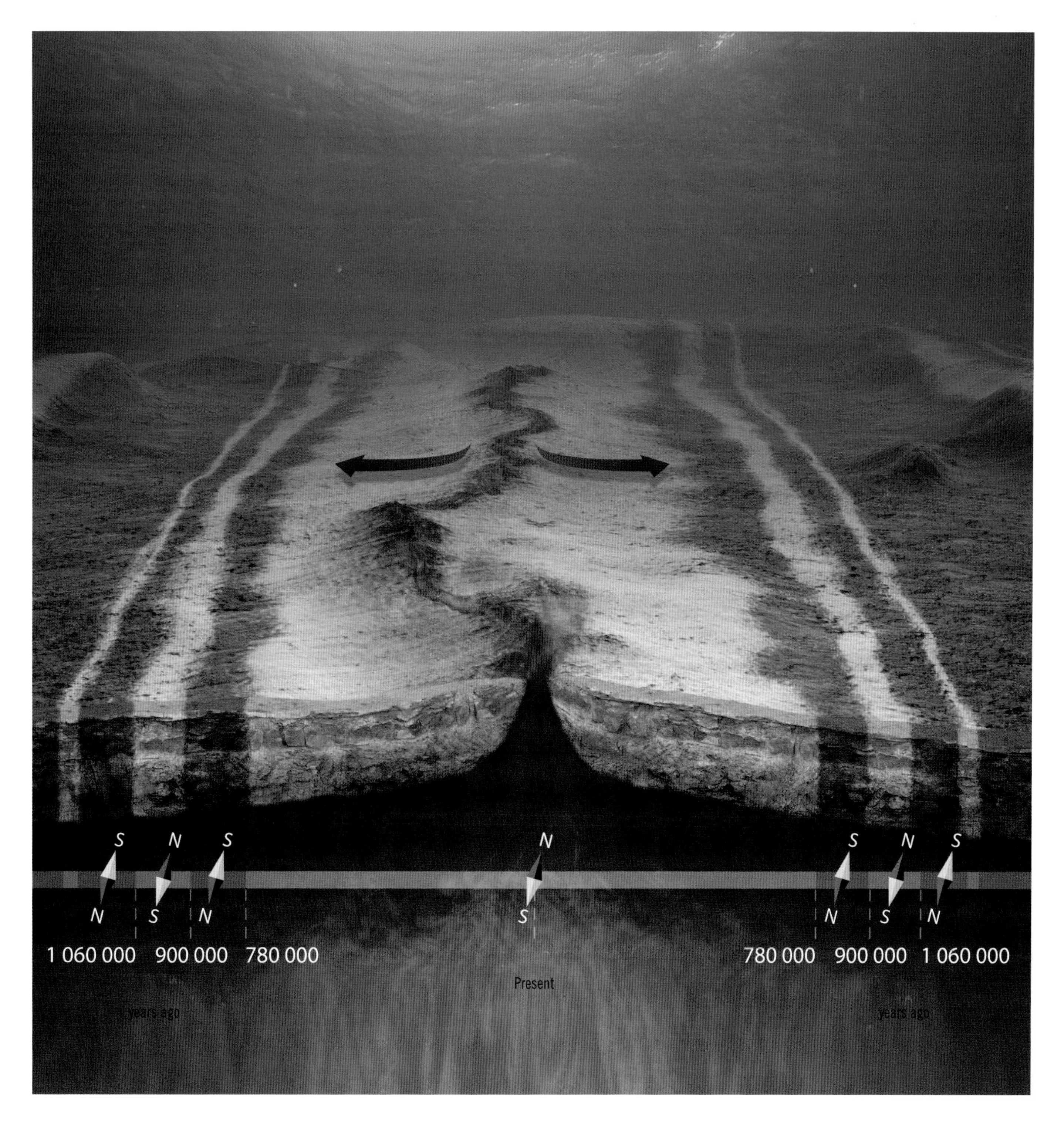

// Ocean trenches

The oceans' most extreme depths are found in trenches—long, narrow depressions formed as one tectonic plate is pushed beneath another.

When continental plates collide at convergent plate boundaries, the older, denser lithosphere melts or slides underneath the younger, less-dense lithosphere in a process known as subduction. As the plates collide, the sea floor and lithosphere buckle, creating a V-shaped depression called a trench. Oceanic trenches are typically about 2 or 2.5 miles (3 or 4 kilometers) deep.

In some cases, the two plates are both carrying oceanic crust, but more often, one carries continental crust and the other carries oceanic crust. Continental crust is more buoyant than oceanic crust, so oceanic crust will always be subducted when these two types of plates meet.

Subduction zones are extremely active seismically, generating some of the largest earthquakes on record. The 2004 Indian Ocean tsunami and the 2011 Tohoku earthquake and tsunami in Japan were both caused by seafloor earthquakes that took place in subduction zones.

Trenches are present in all of the world's ocean basins. They are most common in the Pacific Ocean basin, but there are also examples in the eastern Indian Ocean, Atlantic Ocean and Mediterranean Sea. Globally, there are about 31,000 miles (50,000 kilometers) of convergent plate margins at which about 1 square mile (3 square kilometers) of oceanic lithosphere is subducted each year. At these boundaries there are more than 50 major ocean trenches, with a total area of 734,000 square miles (1.9 million square kilometers).

The deepest trenches are found around the rim of the Pacific Ocean basin as part of the so-called 'Ring of Fire'. Deepest of all is the 1,580-mile-long (2,540-kilometer)

Below *This cutaway diagram of an ocean trench shows how the trench forms at the point where one block of oceanic crust is being subducted beneath another. To the left, an island arc has formed as the crust is buckled and material from the subducted plate rises to form seamounts. To the right can be seen a chain of seamounts rising from the ocean floor.*

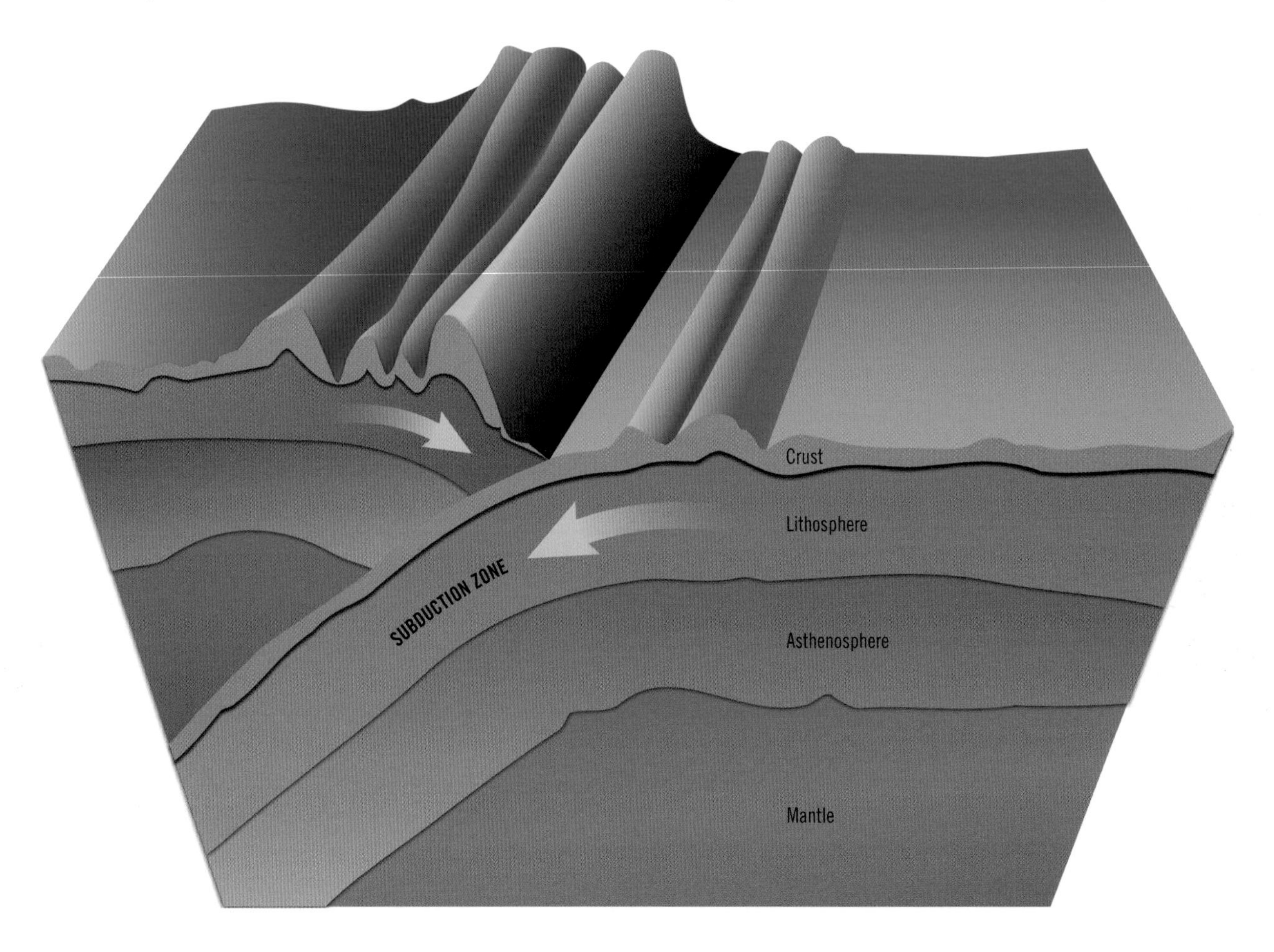

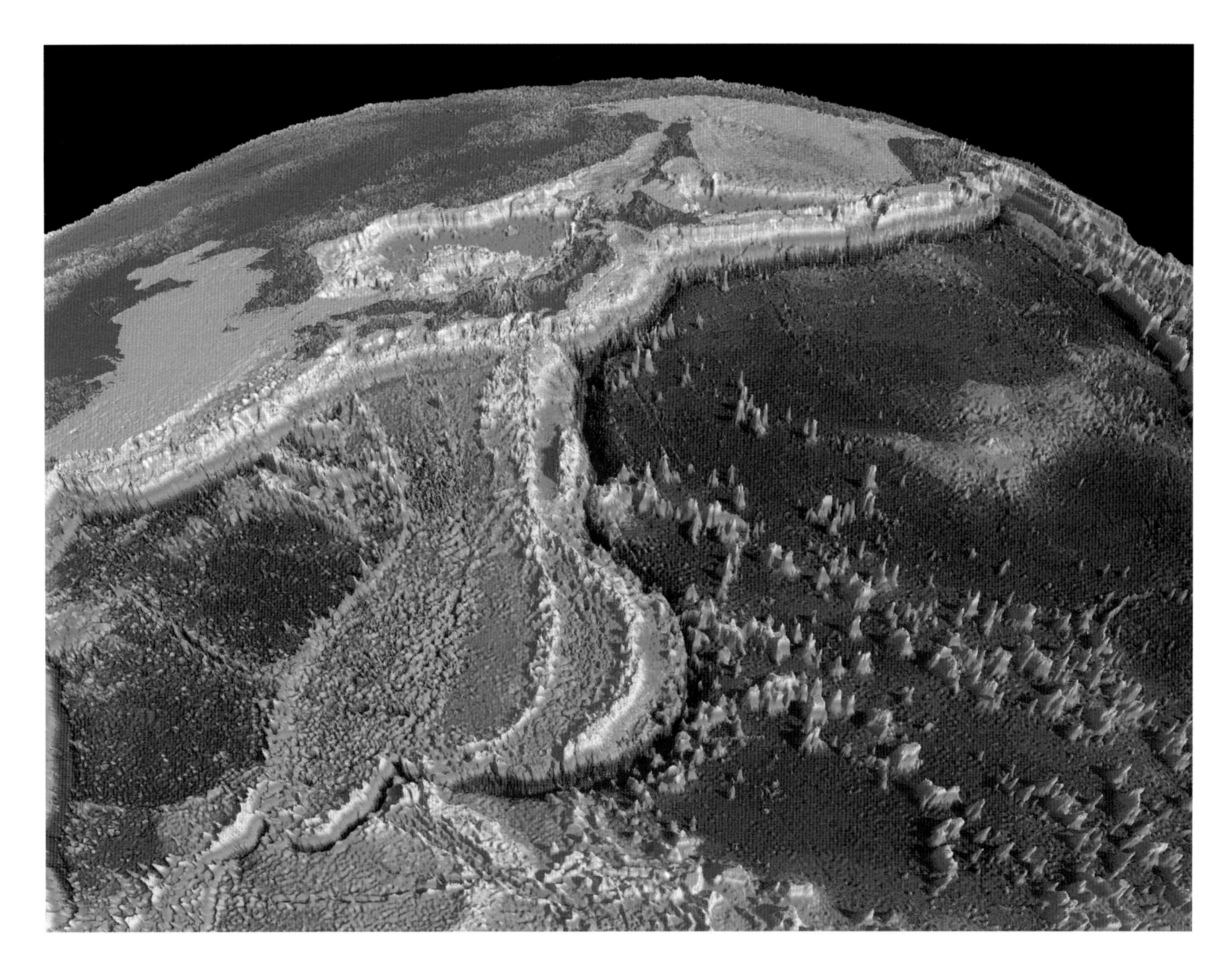

Above *A computer model of the topography of the region surrounding the Mariana Trench, which can be seen as a purple arc in the lower center. The deepest oceanic trench on Earth, the depression forms part of a subduction system in which the western edge of the Pacific Plate is being forced under the smaller Mariana Plate.*

Mariana Trench, which is located near the Mariana Islands in the Philippines. This is an oceanic–oceanic convergent boundary, where the Pacific Plate is being subducted beneath the Philippine Plate. It is home to the Challenger Deep, which, at almost 36,000 feet (11,000 meters) below sea level, is the oceans' deepest point; it is so deep that Mount Everest could fit inside it and still have 1 mile (1.6 kilometers) of water above.

As a plate is subducted, it melts and there is an upwelling of molten crust under the overriding plate that eventually solidifies to form mountain ridges and volcanoes that run parallel to the trench. Molten material flowing up through the volcanoes often forms island arcs, as can be seen in the Japanese archipelago and the Aleutian Islands. These arcs are usually found about 125 miles (200 kilometers) from the trench itself.

As the dense, subducting tectonic plate is pushed downwards, sediment on its surface is sometimes scraped off onto the less-dense, overlying plate. This sediment forms a roughly triangular feature known as an accretionary wedge at the bottom of the trench. If the trench is close to the mouth of a river or glacier, more sediment may flow into it and may even completely fill or overfill it. In the latter case, an island may form, effectively hiding the trench tht lies below. The Caribbean islands of Trinidad and Barbados are examples of this phenomenon; both lie atop the trench created as the South American plate is subducted beneath the Caribbean plate.

The depth of an ocean trench is mostly determined by the rate at which sediment falls into it. Hence the oceans' deepest trenches are found in areas where there is little inflow of sediment.

// The Coriolis effect

The Earth's rotation has a curious influence on both the atmosphere and the oceans, deflecting air and water so that they appear to follow a curving path rather than a straight line. This is known as the Coriolis effect.

Named after the French mathematician and physicist Gustave-Gaspard de Coriolis (1792–1843), the Coriolis effect governs the way in which fluids behave on the Earth.

As the Earth spins on its axis, different regions move at different speeds depending on their distance from the poles. Imagine you are standing at the North Pole and take a step 3 feet away. Because it takes 24 hours for the Earth to complete a rotation, it will take you 24 hours to move around in a circle with a circumference of about 20 feet (6 meters)—so you are moving at about 0.00016 miles (0.00026 kilometers) per hour. Now imagine you're on the Equator. In the same 24 hours, you now travel around the entire circumference of the Earth: about 25,000 miles (40,000 kilometers). Hence, you are now travelling at 1,000 miles (1,600 kilometers) per hour—about six million times faster.

If we now take an airborne object that is travelling north from the equator at a steady velocity, the ground beneath it will be moving from west to east. At first, the ground and the object are at the same speed relative to each other, but as the object travels farther north, the velocity of the ground will slow down relative to it. This has the effect of making the path of our flying object appear to curve to the right (east). Similarly, if the object is heading south, the ground beneath will be speeding up relative to it, so again, it will appear to be heading to the right (west). These curving paths are caused by the Coriolis effect.

If the Earth did not rotate, the atmosphere would simply circulate between the high-pressure areas at the poles and the low-pressure areas at the equator. Instead, air is deflected to the right in the Northern Hemisphere and to the left in the Southern Hemisphere. This means that winds blow in an anticlockwise direction around Northern Hemisphere low-pressure systems and clockwise around high-pressure systems, and vice versa in the Southern Hemisphere (see page 106). The Coriolis effect also has a comparable influence over ocean currents (see below).

The impact of the Coriolis effect depends on the velocity of whatever is being deflected and also the distance that it travels—higher speeds and greater distances equate to more significant deflections. Latitude plays a part too: the Coriolis effect is at its strongest near the poles and absent at the equator.

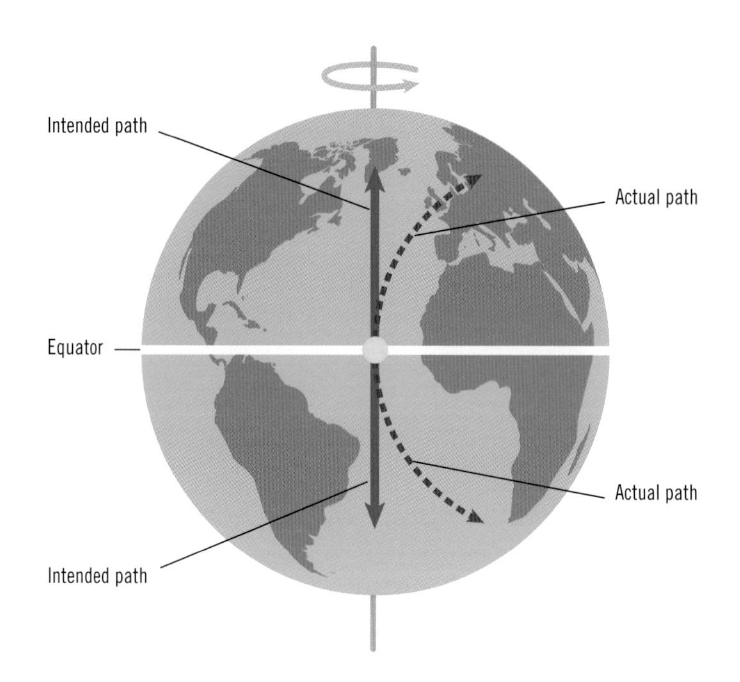

Above *The Coriolis effect causes moving air to be deflected to the right of its intended path in the Northern Hemisphere and to the left in the Southern Hemisphere.*

Ekman transport

The Coriolis effect also has an impact on ocean currents. As wind blows across the ocean surface (1), friction causes it to push the water along (2). Due to the Coriolis effect, the water moves at an angle to the wind direction (3)—about 45° to the right in the Northern Hemisphere and to the left in the Southern Hemisphere. This angle becomes more pronounced with increasing depth, leading to the creation of a spiral of moving water that is 330–500 feet (100–150 meters) deep, known as an Ekman spiral, after the Swedish scientist V. Walfrid Ekman (1874–1954), who first described it in 1905.

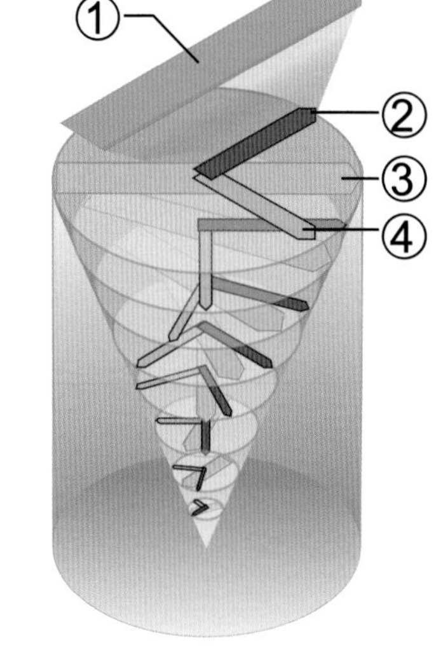

Right *Ekman spiral.*

On average, the direction of movement of the water in an Ekman spiral is at roughly a right angle to the wind direction (4); this is called Ekman transport. The layer of water that is affected by the movement of wind-driven surface waters typically extends to a depth of about 330 feet (100 meters) and is known as the Ekman layer.

In shallow water, because the water depth is insufficient for a full Ekman spiral to develop, the angle between the wind direction and surface-water movement is reduced, in some cases to as little as 15°.

Above *A low-pressure system over Iceland showing the inward anti-clockwise spiraling of the winds.*

// Ocean currents

Flowing like vast rivers in the ocean, currents move enormous amounts of water around the planet.

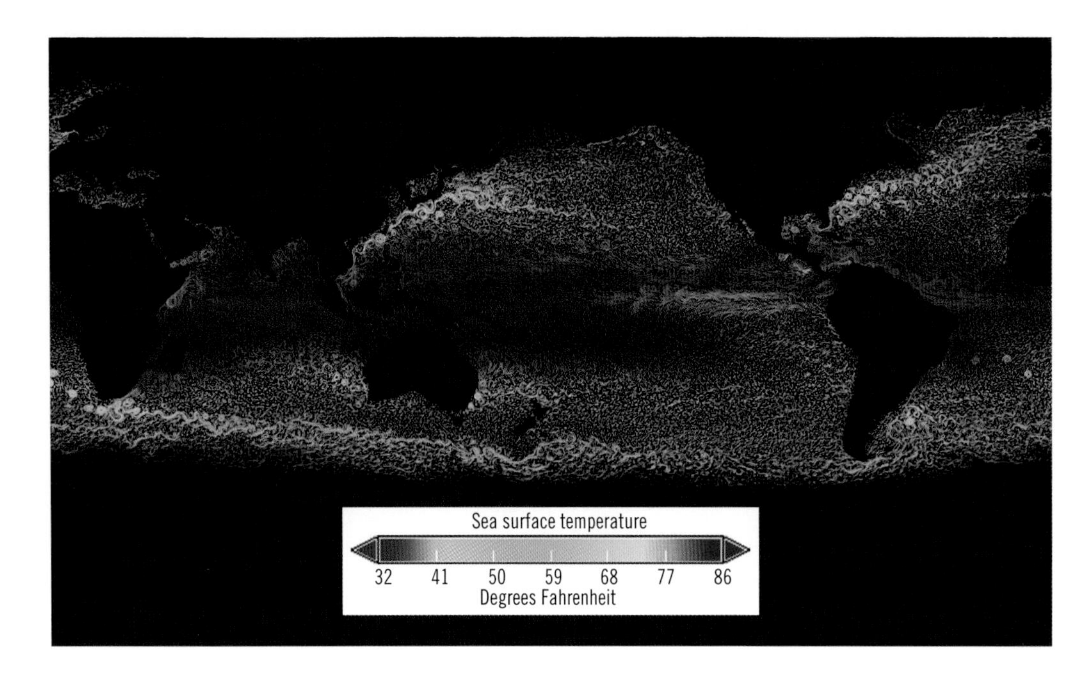

Left *Global ocean current flows color-coded according to the sea-surface temperature.*

In each of the world's ocean basins, surface currents join together to form gyres, permanent circular currents that are thousands of kilometers in diameter (see box).

Current velocities range from a few centimeters per second to as much as 13 feet (4 meters) per second. Wind-driven surface currents tend to move significantly more quickly than those associated with the thermohaline circulation: about 20 inches (50 centimeters) per second compared to 0. 4 inches (1 centimeter) per second. The Gulf Stream moves relatively rapidly: along the coast of North America, it reaches a speed of about 100 inches (250 centimeters) per second (equivalent to 6mph or 9km/h) at the sea surface.

The intensity of a surface current generally decreases with increasing depth as the wind's influence fades and the density of seawater increases. The depth penetration depends on how stratified the water is: in regions such as the tropics where the water is strongly stratified, currents are mostly restricted to depths of less than 3,300 feet (1,000 meters), but towards the poles, where stratification is weaker, the wind-driven circulation generally reaches all the way to the sea floor.

Surface currents play a significant role in global climate and local weather patterns by transferring large amounts of heat from equatorial areas to the poles (see page 114). The Gulf Stream carries about 150 times more water than the Amazon River. It keeps much of Northern Europe warmer than other places at similar latitudes.

Ocean currents are measured in sverdrups: one sverdrup is equivalent to a flow rate of 35 million cubic feet (1 million cubic meters) of water per second. Taken together, surface currents move about 8 percent of the ocean's water.

Continuous and directed movements of seawater, ocean currents can be long, permanent features, such as the Gulf Stream, or shorter, episodic flows, such as the shallow longshore currents that run along coastlines. They can be divided into surface currents, which are mostly driven by wind systems, and deep-ocean currents, which are mostly driven by differences in water density due to differences in temperature and salinity (so-called thermohaline circulation; see page 110). There are also tidal currents, which flow as the ocean rises and falls with the tides. These tend to be strongest when water is forced through a narrow gap.

The direction and strength of an ocean current are influenced by numerous factors, including the wind, the Coriolis effect, ocean-basin and shoreline topography, and temperature and salinity differences.

The world's large surface currents are driven by the prevailing winds. As the winds blow over the ocean's surface, frictional drag moves the water in the wind's direction. The Coriolis effect and resulting Ekman transport (see page 106) then cause the current to bend to the left or right, depending on the hemisphere in which it is flowing.

As ocean currents flow, they sometimes form eddies—relatively small contained loops of swirling water that break off from the current's main body and travel independently, sometimes for long distances, before dissipating.

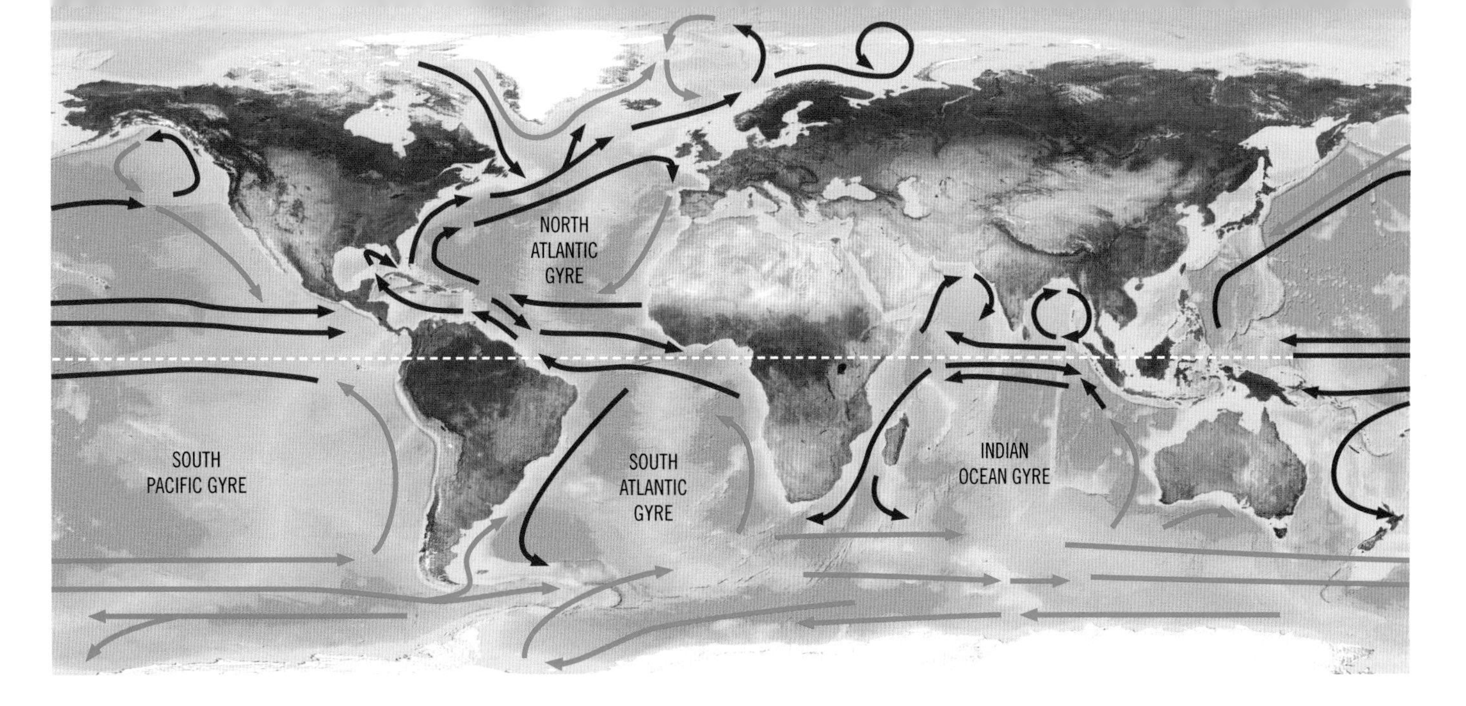

OCEAN GYRES

Thousands of miles in diameter, ocean gyres are made up of numerous currents that link together to form a permanent circular current that follows the coastlines of the Earth's continents. They are driven by three factors: global wind patterns, the Earth's rotation, and the Earth's landmasses. Wind pushes on the surface water, causing it to move. The Earth's rotation then deflects the direction of the water's movement through the Coriolis effect. Landmasses also deflect the water, forming a boundary around the outside of the gyre. Because of the Coriolis effect, gyres tend to turn a clockwise direction in the Northern Hemisphere and counter-clockwise in the Southern Hemisphere.

In ocean gyres, the western boundary current is the deeper, warmer, stronger, and narrower of the north–south currents. These currents, which include the Gulf Stream in the North Atlantic, the Kuroshio in the North Pacific and the Agulhas in the Indian Ocean, are among the world's fastest non-tidal ocean currents, reaching speeds of up to 4.3 mph (7 km/h) and containing as much as 100 times the combined flow of all the world's rivers. The tropical gyres, which operate near the Equator, tend to be less circular, flowing mostly east to west and back.

Most ocean gyres are stable and predictable; for example, the North Atlantic Ocean Gyre always flows in a steady clockwise path around the North Atlantic Ocean. However, some experience seasonal variation. The flow of the Indian Ocean Gyre, for instance, is governed by the monsoonal weather system. In the summer, it flows in a clockwise direction, as the prevailing winds blow in from the ocean, while in the winter it flows counter-clockwise as the wind blows in from the Tibetan plateau.

Above *The world's main ocean gyres. Warm-water currents are marked in red; cool-water currents are marked in blue.*

Below *The pattern of ocean currents in any particular region is often quite complex. For example, four major currents flow through Australian waters: the East Australian Current, which flows south along Australia's east coast; the Leeuwin Current, which flows from mid-way down the Western Australian coast around to the west coast of Tasmania; the Antarctic Circumpolar Current, which flows west to east south of Australia; and the Indonesian Throughflow, which brings warm water from the Pacific to the Indian Ocean via Indonesia.*

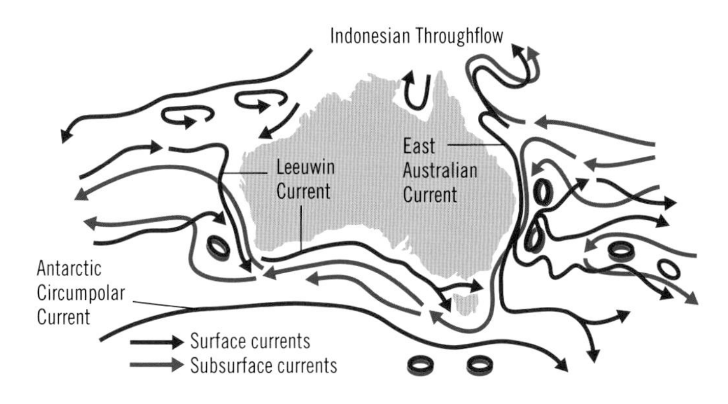

THE ANTARCTIC CIRCUMPOLAR CURRENT

The near-constant westerly winds that blow over the Southern Ocean drive a powerful ocean current that encircles Antarctica. Known as the Antarctic Circumpolar Current, it flows from west to east, transporting 424 million cubic feet (125 million cubic meters) of seawater per second over a path of about 15,000 miles (24,000 kilometers). Moving about twice the volume of the Gulf Stream, it reaches the sea floor and is guided along its course by the irregular bottom topography. The current is responsible for the large upwelling of cold water that exists around the Antarctic and helps to drive the ocean conveyor belt.

// The ocean conveyor belt

Ocean water is in constant motion, driven around the globe by a planet-spanning circulation pattern known as the ocean conveyor belt.

The pattern of currents that forms the ocean conveyor belt is also called thermohaline circulation as it is primarily driven by two factors that determine the density of water: temperature (thermo-) and salinity (-haline).

The "beginning" of the ocean conveyor is located in the Norwegian Sea, which is part of the North Atlantic. Here, warm, salty water from the Gulf Stream heats the atmosphere in the cold northern latitudes and itself becomes cooler and thus denser. Its density increases further when sea ice forms; salt doesn't freeze, so it remains in the water. This water then begins to sink, allowing more warm water to take its place.

Down in the deep ocean, this water mass, known as the North Atlantic Deep Water (NADW), moves southwards, into the deep abyssal plains of the Atlantic, past the Equator and on to the Antarctic, where it is brought back to the surface by the region of upwelling that encircles the Antarctic continent. Some of this water then begins to travel north on the surface, eventually making its way into the Gulf Stream, which runs up through the Atlantic past Europe. Although the Gulf Stream is primarily a wind-driven current, thermohaline circulation contributes about 20 percent of its flow.

Around Antarctica, the conveyor is also recharged, as strong winds cool surface water and the formation of sea ice increases salinity. The resulting Antarctic Bottom Water is extremely dense and, as it flows to the north and east, it moves beneath the NADW.

The deep-water portions of the conveyor belt move relatively slowly, just a few centimeters per second or less, compared to wind-driven or tidal currents, which can move

Below *The ocean conveyor "begins" in the North Atlantic, as cold, salty water (blue) sinks into the deep ocean. It then moves southwards, past the Equator and on to the Antarctic, where it splits into two streams, one of which moves into the Indian Ocean, while the other moves into the Pacific. Both eventually return to the surface (red) and then re-join to form the Gulf Stream, which runs up through the Atlantic past Europe and into the North Atlantic to complete the cycle.*

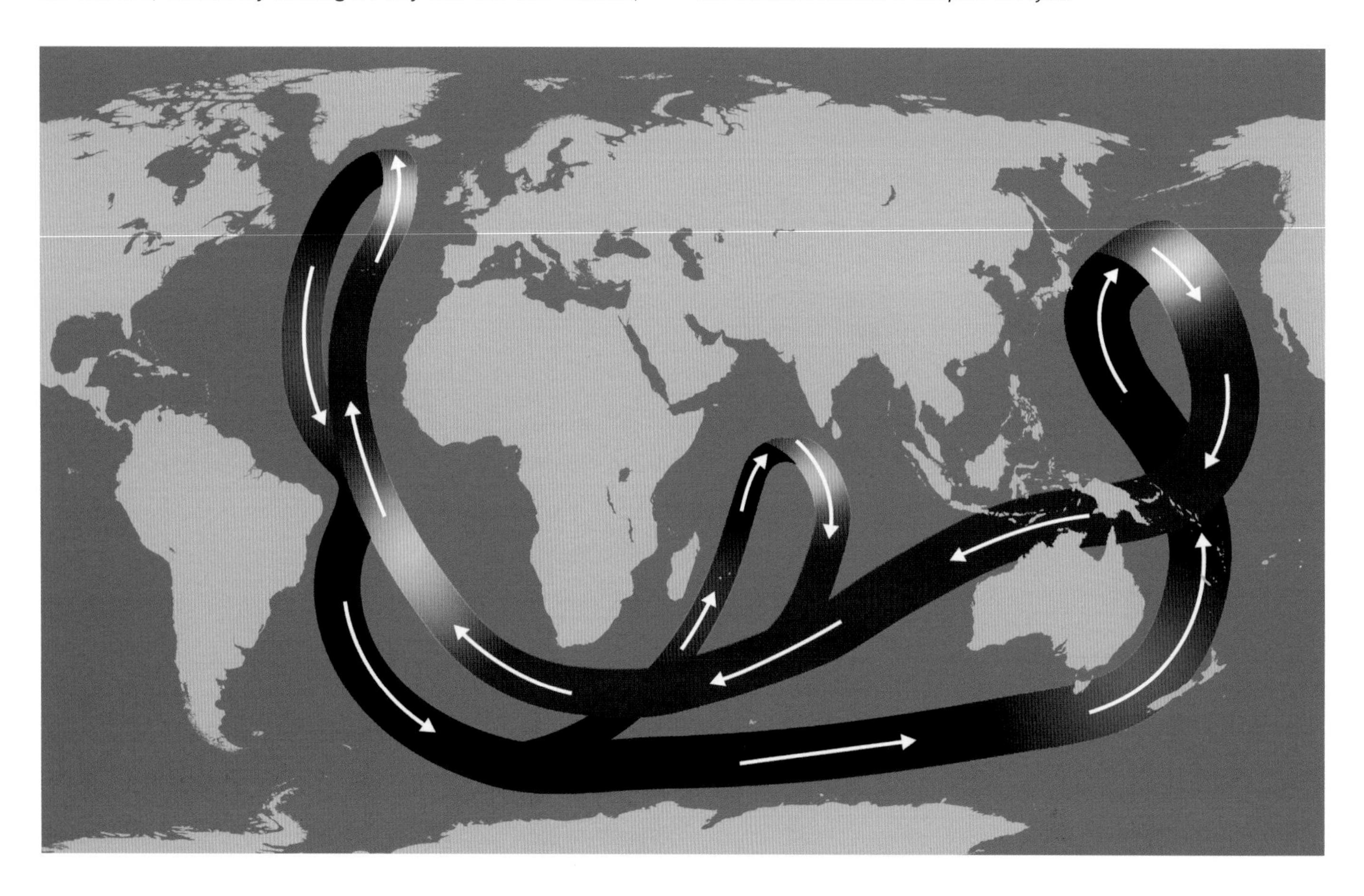

at tens or even hundreds of inches per second. It has been estimated that water takes about 1,000 years to make its way all the way around the conveyor belt. The total volume of water moving around the conveyor amounts to about 96,000 cubic miles (400,000 cubic kilometers), around a third of all the water in the oceans.

By effectively moving heat absorbed from the Sun around the globe, the conveyor belt plays a crucial role in shaping the Earth's climate. The warm water that flows into the Arctic in the Gulf Stream regulates the growth of sea ice, which has an impact on the absorption of solar energy by changing the albedo (reflectivity). It also has an impact on the local climate, keeping adjacent regions warmer than they would otherwise be; thanks to the Gulf Stream, the average annual temperature in Norway is nearly 18°F (10°C) warmer than that in Manitoba, Canada, which is at a similar latitude.

The conveyor belt also drives the cycling of nutrients and carbon dioxide around the ocean. Warm water at the ocean's surface is typically low in nutrients and contains little carbon dioxide, but as it cools it takes up more carbon dioxide and, as the conveyor transports it through the deep ocean, it picks up nutrients. This nutrient-rich water is brought to the surface at upwellings (see page 112), leading to the creation of extremely productive regions.

Below *Evaporation, cooling and the freezing of sea ice increase the salinity—and hence the density—of surface water, causing it to sink. The shape of the ocean basins then forces this cold, dense water to flow south, before it eventually returns to the surface via upwellings.*

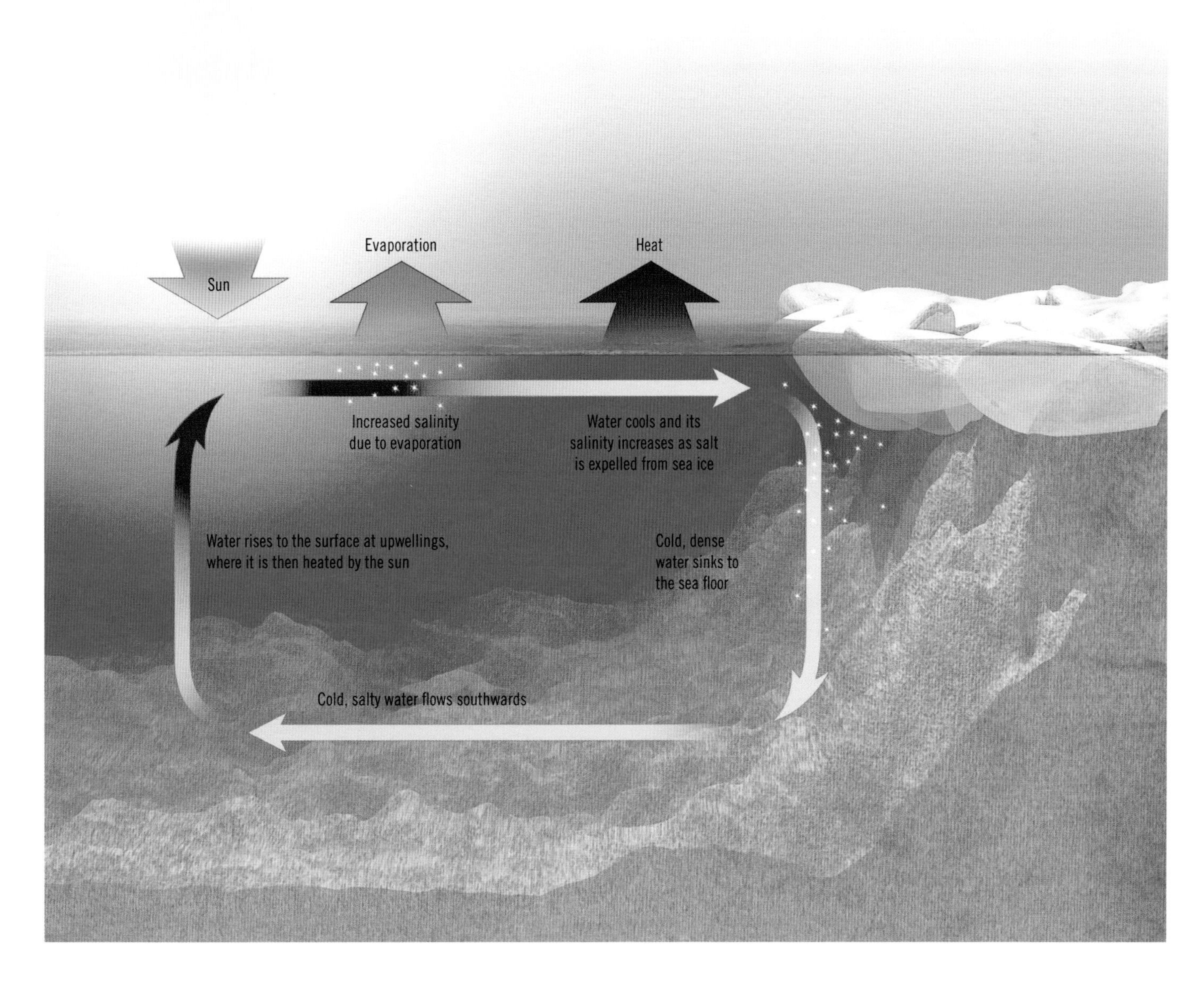

// Upwellings

Scattered around the world's oceans are regions where cold, nutrient-rich water is drawn up from the ocean depths to the surface. Known as upwellings, these areas are highly productive, supporting rich assemblages of marine life, and can also influence global climate.

Upwellings can be found both in the open ocean and along coastlines. They are mostly driven by surface winds. As these blow surface water away, deeper water rises up to replace it. In coastal regions, optimal conditions for upwelling occur when winds blow along the shore. The Coriolis effect and associated Ekman transport then create currents that flow out to sea.

The reverse process, known as downwelling, occurs when onshore winds force warm surface water against the coastline, causing it to sink. In some regions, such as the west coast of the USA, upwelling and downwelling patterns alternate seasonally as the prevailing winds change direction. Most downwelling occurs in subtropical waters. Largely devoid of nutrients, these areas have little in the way of marine life.

An upwelling may also occur when the topography of the sea floor, such as that around islands, ridges, and seamounts, causes deep-ocean currents to flow upwards; examples include upwellings around the Galápagos Islands and the Seychelles. Slow-moving cyclones, too, can create short-lived upwellings. Their winds blow surface water aside, causing upwelling to take place beneath the eye of the cyclone. The colder water that is drawn up eventually weakens the cyclone.

There are regions of upwelling just north and south of the equator, caused by the easterly trade winds. The Coriolis effect sends surface waters north and south, resulting in an upwelling that makes the equatorial region in the Pacific visible from space as a broad line of high phytoplankton concentration.

In the Southern Ocean, the strong westerly winds that blow around Antarctica drive surface water northwards, in the process drawing large quantities of cold water to the surface in a ring that encircles the Antarctic continent; research suggests that as much as 80 percent of the deep water in the oceans resurfaces in the Southern Ocean. Much of this water was last exposed to the air as much as 1,000 years previously; most of it is drawn up from a depth of about 6,600–10,000 feet (2,000–3,000 meters).

While upwellings are mostly localized phenomena, they can have wide-ranging impacts. For example, upwellings off the west coast of South America play a central role in the El Niño-Southern Oscillation (see page 164), which can alter temperature and rainfall patterns across a very broad swathe of territory.

The area of an upwelling can be extremely large; the upwelling off the west coast of Peru covers about 10,000 square miles (26,000 square kilometers) of ocean.

Nutrients brought up with the cold water—mainly nitrogen and phosphorus from the decomposing bodies of dead sea creatures—effectively act as ocean fertilizers, stimulating the growth and reproduction of phytoplankton. These phytoplankton are consumed by zooplankton, which are consumed by larger marine animals. Hence upwellings often support thriving fishing industries; about half of the total global marine fish catch comes from coastal upwelling zones, which cover only 1 percent of the total area of the world's oceans.

Below *Coastal upwellings are typically driven by winds blowing parallel to the shoreline. The Coriolis effect and associated Ekman transport create surface currents that cause water to flow out to sea and deeper water rises up to replace it.*

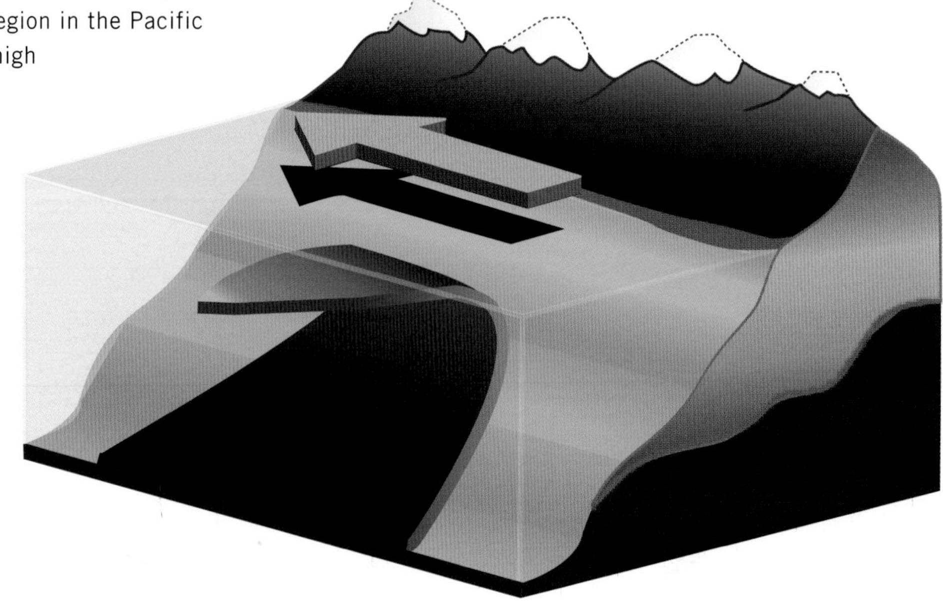

A satellite image of phytoplankton associated with an upwelling off the east coast of the USA. Nutrients drawn up to the surface fertilize the tiny plants, creating a highly productive ecosystem.

// Ocean heat and nutrient transport

The oceans are a global storehouse for heat and help to regulate the Earth's climate by transporting some of that heat from the equator to the poles. Ocean circulation also moves dissolved nutrients around, helping to determine where marine life is found.

On average, each year in the tropics and subtropics, more heat arrives at the Earth's surface than is released. From the higher latitudes to the poles, the opposite is true. Hence the atmosphere and the oceans move energy from the equator towards the poles in order to compensate for this imbalance.

The amount and direction of heat transport varies between the different ocean basins. In the Pacific Ocean, heat is transferred to the north and south from the equator, while there is very little northward heat transport in the Indian Ocean. Only in the Atlantic Ocean does heat travel all the way from south to north—across the equator. In general, heat transport from one ocean basin to another is relatively minor.

Global ocean heat transport is dominated by the export of heat from the tropical Pacific; in some parts of the tropical eastern Pacific, the ocean acquiresnearly 10 watts of heat per square foot (more than 100 watts per square meter). Most of the heat is lost in the North Atlantic, North Pacific, and Arctic oceans, particularly off the eastern coasts of North America and Asia and in parts of the Arctic, where up to 20 watts per square foot (200 watts per square meter) are transferred to the atmosphere.

In the Atlantic, most of this heat transfer is driven by the ocean conveyor belt, rather than by wind-driven currents. In the Pacific and Indian oceans, most of the heat transport is carried out by the surface wind-driven gyres.

Taken together, ocean currents transport just under 3 petawatts of heat to the north; this represents about 600 times the energy produced by all of the world's power stations. Atmospheric circulation adds another 2.5–3 petawatts of heat, resulting in a total northward transport of 5.5–6 petawatts.

Warm currents increase the local air temperature by warming the sea breezes that blow over them and vice versa for cold currents. Europe enjoys a warmer climate because of the heat carried north by the Gulf Stream and the North Atlantic Current; parts of Canada at a similar latitude are typically 18°F (10°C) cooler in winter because they lack this warming effect. If the Gulf Stream were to shut down, the annual mean temperature in northern Norway would

Below *This visualization shows sea-surface current flows, colored to match the corresponding sea-surface temperatures. It illustrates the ocean eddies and other narrow current systems that transport heat and carbon around the oceans.*

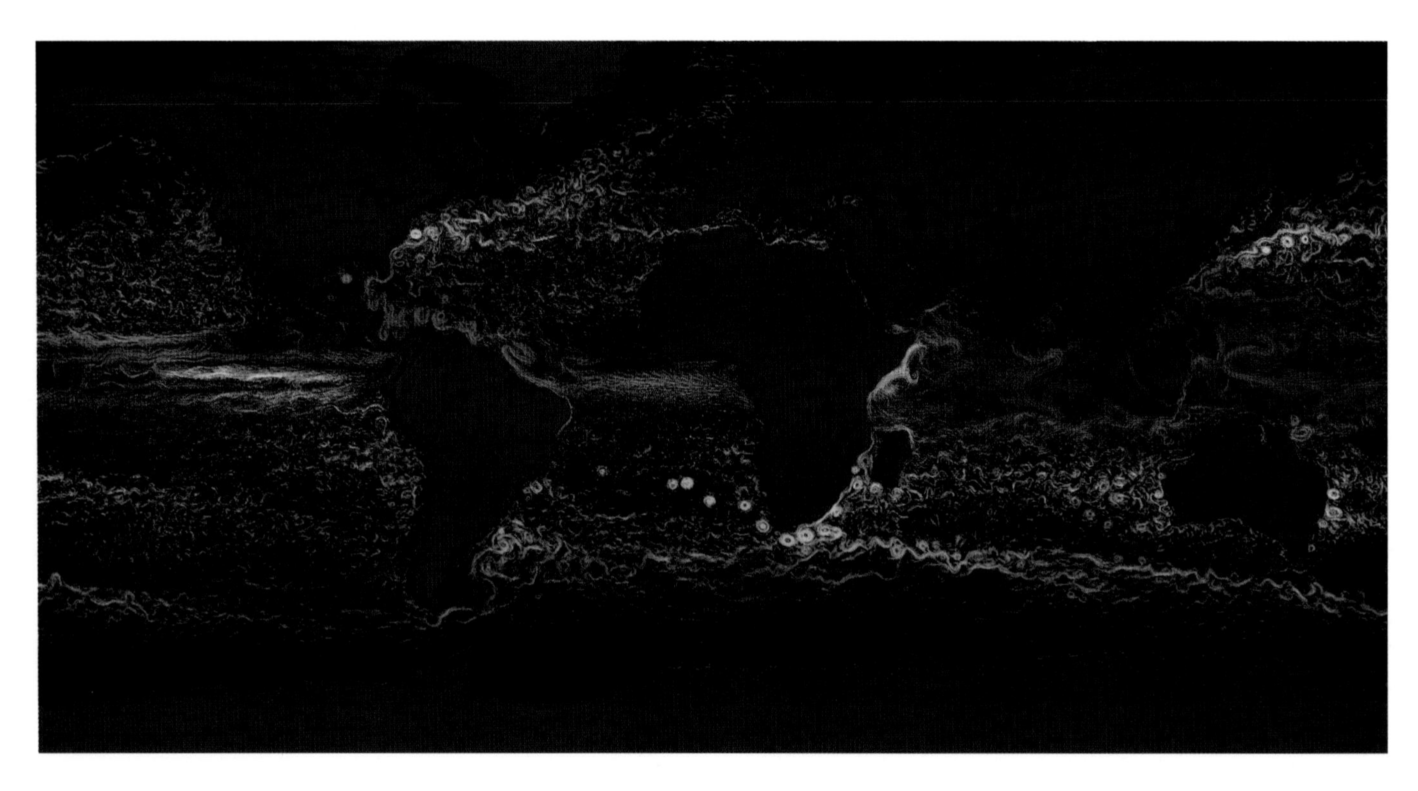

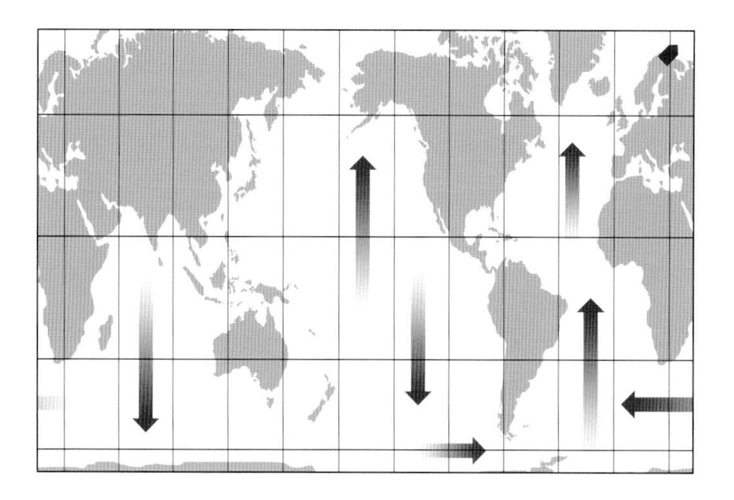

Above *In the Pacific Ocean, heat travels north and south from the equator; in the Indian Ocean it only travels south; in the Atlantic, it travels from south to north along the whole of the ocean basin.*

decrease by more than 27°F (15°C). Conversely, Lima, Peru, is cooler than cities at similar latitudes elsewhere because of the effect of the cold Humboldt Current.

The arrival of warm water in the Arctic regulates global temperature through its effects on the growth of sea ice. Warm currents can also influence rainfall and the formation of storms, providing the moisture and heat that fuel clouds and cyclones.

Nutrient transport

A nutrient is an element that organisms require for maintaining life and growth. The most important nutrients in seawater are nitrogen, phosphorus, and silicon, although trace metals, notably iron, also play a significant role.

Rivers are the main source of the ocean's nutrients. At the ocean's surface, nutrients are consumed by phytoplankton and cyanobacteria. When these and other organisms die, their bodies sink, so the decomposition of those bodies and the release of nutrients back into the seawater mostly takes place in deeper waters. As a result, nutrient concentrations are usually low near the surface and greater near the sea floor. The exceptions to this rule are found at upwellings (see page 112), where cold, nutrient-rich seawater rises to the surface, and over the continental shelf (see page 94), where nutrient levels are boosted by river discharge and wind and tidal mixing, which can bring nutrients back up to the surface.

Large eddies and ocean gyres sometimes sweep nutrient-rich water from the continental shelf and into the open ocean; however, much of this water is carried out below the euphotic zone, so the nutrients are not available to the local marine life.

Below *This satellite image shows the "brightness temperature" observed at the top of the atmosphere above the Gulf Stream in the Atlantic Ocean—that is, the heat radiation from a combination of the sea surface and the overlying moist atmosphere.*

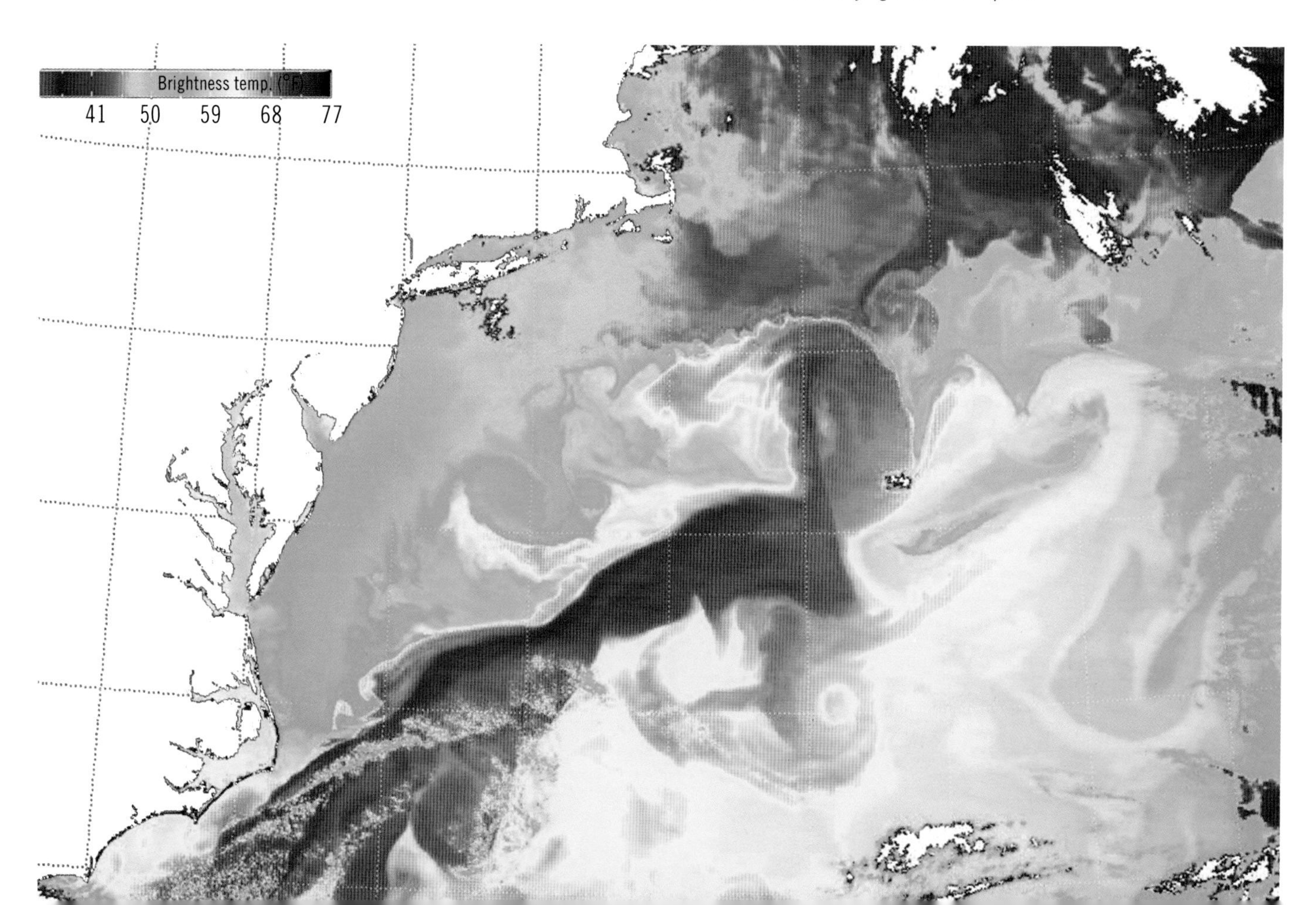

// The ocean carbon cycle

The oceans are the Earth's largest storehouse of carbon, another way in which they help to regulate global climate.

Oceanic carbon can be broken into four distinct categories depending on whether it is organic or inorganic and dissolved or particulate. Organic carbon is found in organic compounds such as proteins and carbohydrates, while inorganic carbon is mostly found as simple compounds such as carbon dioxide and bicarbonate ions.

The oceanic carbon cycle moves both organic and inorganic carbon around the oceans and exchanges it between the atmosphere, the Earth's interior and the sea floor. The main sources of input to the marine carbon cycle are the atmosphere and rivers. Each year, about 200 million tons (180 million tonnes) of carbon is transported to the oceans by rivers.

Three main processes, known as pumps, drive the cycle: the solubility pump, where atmospheric carbon dioxide dissolves in water; the carbonate pump, where marine organisms take up carbon when they make their calcium carbonate shells; and the biological pump, where phytoplankton and other plants turn carbon dioxide into organic matter through photosynthesis.

Carbon dioxide dissolves more easily in cold water than in warm, so the solubility pump is more active at the poles. There, water with a high carbon dioxide concentration sinks and begins traveling in the ocean conveyor belt (see page 110). When it returns to the surface at upwellings (see page 112), the water warms and the carbon dioxide returns to the atmosphere. The rate at which carbon dioxide is taken up by the ocean also depends on the concentration of carbon dioxide already present in the atmosphere and the ocean, as well as the salinity and the wind speed.

Both the carbonate and biological pumps remove carbon from seawater and turn it into either calcium carbonate or organic matter, reducing the amount of carbon dioxide in the surface waters and allowing more carbon dioxide to dissolve. When the organisms responsible die, some of that material sinks to the ocean floor, where it is either buried or decomposes. In the former case, the organic carbon is effectively taken out of the carbon cycle for a significant time period; calcium carbonate skeletons can suffer the same fate, ultimately being turned into limestone. In the latter case, the carbon is added to the cold bottom water.

Humans have had a significant impact on the marine carbon cycle. Prior to the Industrial Revolution, the ocean acted as a net source of carbon dioxide to the atmosphere. Now, however, the oceans have become the world's largest carbon sink, absorbing some 15–40 percent of human-generated carbon dioxide.

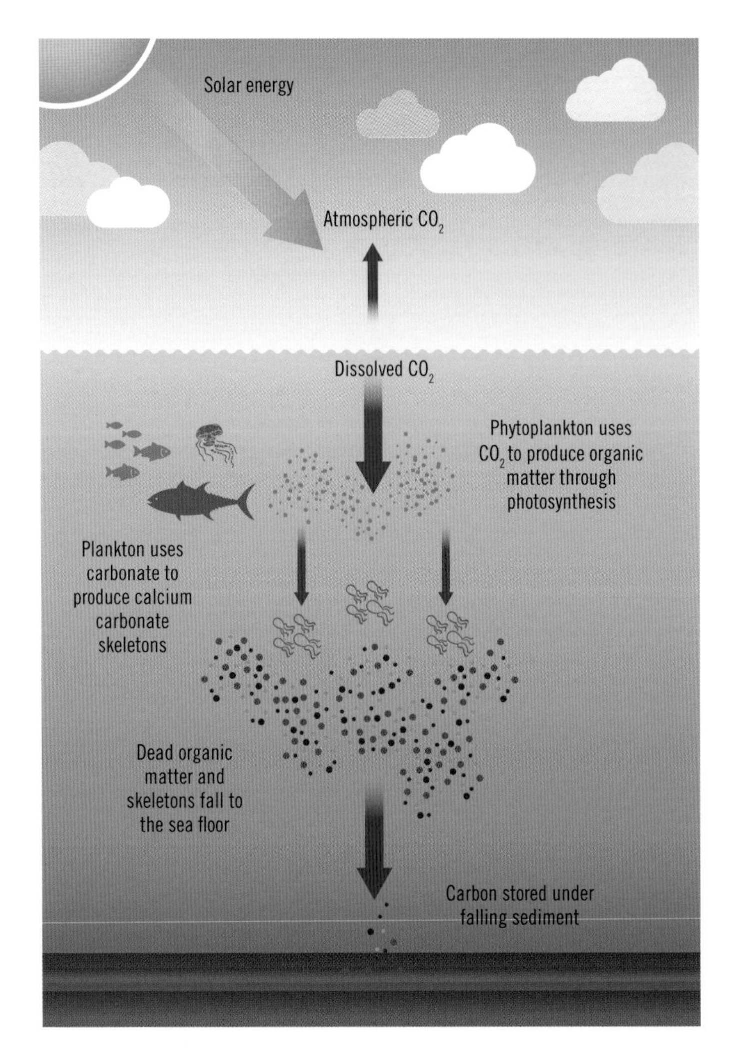

Above *In the ocean's biological pump, dissolved carbon dioxide is turned into organic matter and calcium carbonate skeletons by phytoplankton, coral polyps and other organisms. Much of this carbon is then locked up when it sinks to the seafloor and is covered by sediment.*

There is about 42,000 gigatons (38,000 gigatonnes) of carbon at the Earth's surface, about 95 percent of which is stored in the ocean, mostly as dissolved inorganic carbon. About 1.75×10^{15} kilograms of carbon are tied up in ocean-floor sediment. The ocean holds about 50 times as much carbon as the atmosphere. About 20 gigatons (18 gigatonnes) of carbon taken up through photosynthesis is consumed by other organisms each year.

OCEAN ACIDIFICATION

The increase in the amount of carbon dioxide entering the oceans is affecting the pH of seawater. When carbon dioxide dissolves in water, it becomes a weak acid. The acidification of ocean water reduces the availability of calcium carbonate, which could have a significant negative impact on many skeleton-forming sea creatures, including coccolithophores and foraminifera, which form the base of the marine food chain in many areas, and reef-building corals. If acidification continues to increase, it could also potentially lead to shells and coral skeletons being eroded.

A coral pinnacle in the Red Sea, Egypt. Ocean acidification poses a threat to corals and a range of other marine organisms.

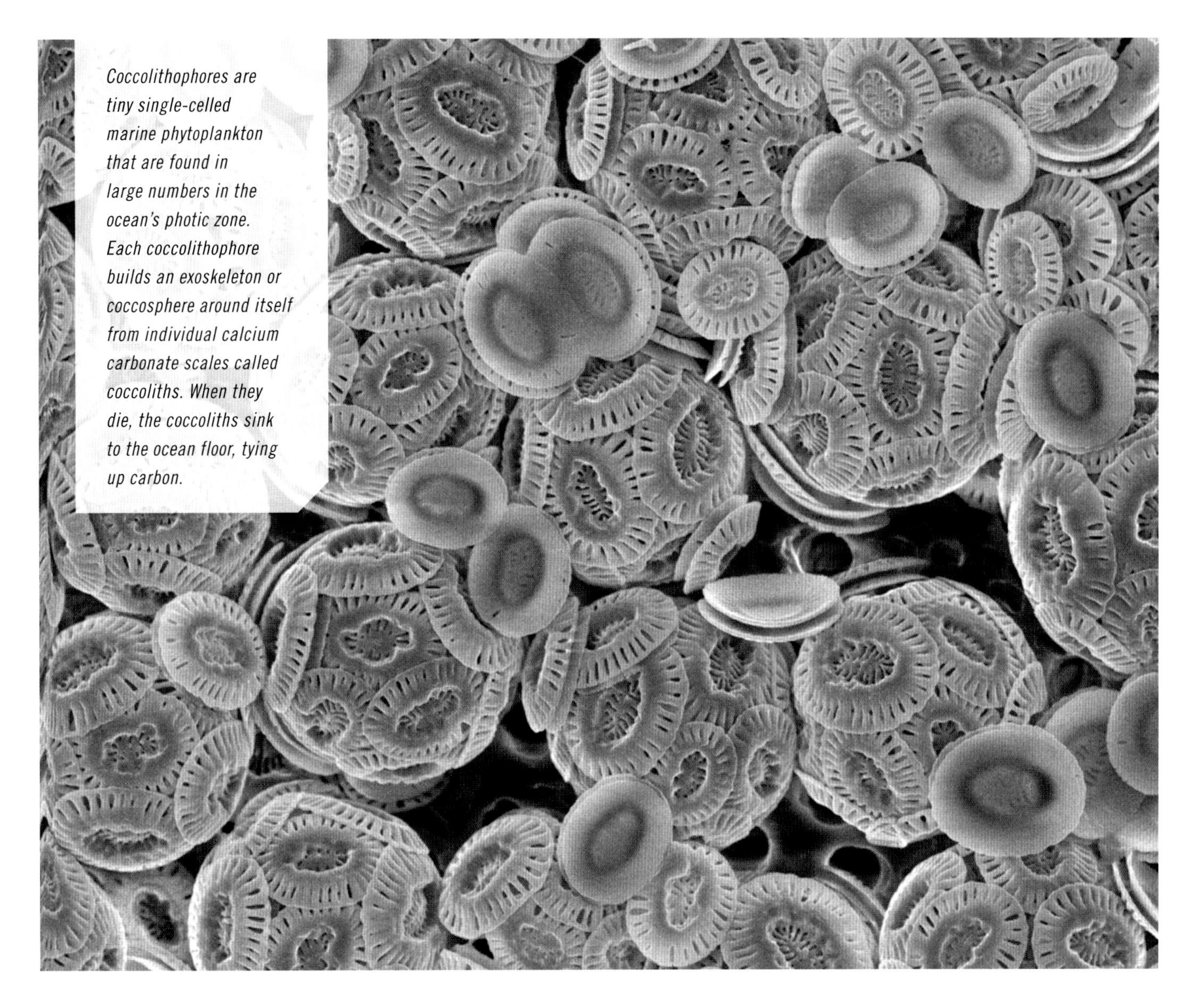

Coccolithophores are tiny single-celled marine phytoplankton that are found in large numbers in the ocean's photic zone. Each coccolithophore builds an exoskeleton or coccosphere around itself from individual calcium carbonate scales called coccoliths. When they die, the coccoliths sink to the ocean floor, tying up carbon.

// Waves

The surface of the ocean is in constant motion, undulating restlessly as waves roll in from sea to shore.

Above *A breaking wave. Ocean waves typically break as they reach shallow water. Friction with the seafloor causes the base of the wave to slow. As the top is still moving at normal speed, it eventually topples over.*

The highest part of a wave is called the crest; the lowest part is the trough. The most important characteristics of waves are their height (the vertical distance from trough to crest), wavelength (the horizontal distance from crest to crest) and period (the time that separates the arrival of consecutive crests at a stationary point). In general, the longer the wavelength, the faster the wave energy will move through the water. Wind waves typically have wavelengths of about 200–500 feet (60–150 meters) and a period of about 20 seconds.

Ocean waves are generally classified according to the energy source that created them. The most common are surface waves, which are caused by wind blowing across the water's surface. Under normal conditions, these are the waves that are seen at a beach.

The waves that roll in to the beach will usually have been generated by winds blowing far away. When wind first starts to blow over an area of water—known as a fetch—friction causes ripples or capillary waves with small wavelengths to form on the surface. As the wind continues to blow, the ripples start to grow into larger waves because the wind has greater purchase on the uneven water surface. The largest surface waves form when strong winds blow steadily for a long time and over a long distance; large waves will not form if strong winds only gust for a short time.

When the waves are large enough, they begin to travel out of the fetch, naturally separating into swells—groups of waves that have a common direction and wavelength.

They then continue across the ocean until they meet the shore. If they do not meet any obstructions, waves can potentially travel across an entire ocean basin; strong winds off Tasmania can create perfect waves for surfing in southern California.

A sea is said to be fully developed when it has the maximum wave size theoretically possible for a given wind strength, duration, and fetch. When wind continues to blow over a fully developed sea, it causes the wave crests to break in what are known as whitecaps.

The advancing waves that we see are actually energy passing through the water, rather than the water itself moving forwards. As a wave's energy passes through, it causes the individual molecules to move up and down in a circular motion.

If a wave is tall enough, it will eventually exhibit a phenomenon known as breaking. Waves break when their base can no longer support their top, causing them to collapse. This can take place in the open ocean, but happens more predictably when waves reach the shore.

SEICHES

Fully or partially enclosed water bodies such as lakes and seas sometimes develop what are known as seiches. A type of standing wave, a seiche forms when a disturbance creates waves that travel across the water body and are then reflected. The waves eventually harmonize so that some of the water moves up and down, while other parts—called nodes—remain stationary.

As waves move into shallow water, their height increases, speed decreases, and wavelength decreases, a process known as shoaling. Friction between the water and the sea floor slows the base of the wave down and the wave's lower portion is also compressed, forcing the crest higher. As the top part of the wave is continuing at the same speed, the crest eventually falls over and crashes down.

Below *Some of the important characteristics of waves.*

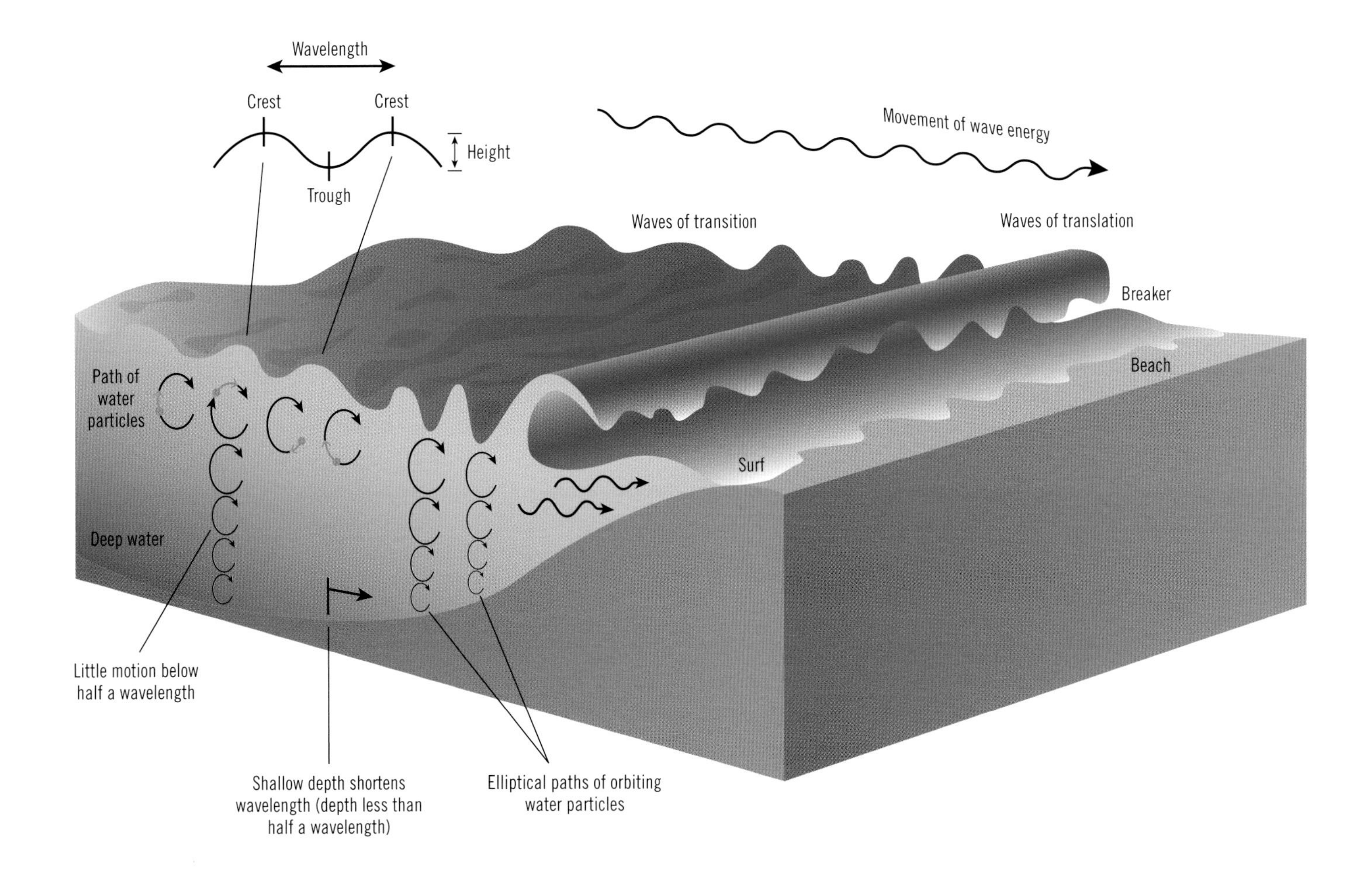

// Tsunamis

Enormous surges of water that are among the most destructive natural phenomena on Earth, tsunamis have caused countless deaths over the centuries.

Tsunamis are generated by the sudden displacement of a large volume of water, typically by a seismic event such as an earthquake or volcanic eruption, but also potentially by a terrestrial or marine landslide, glacier calving or a meteorite impact. Most tsunamis are caused by large undersea earthquakes. When an earthquake makes the ocean floor rise or fall suddenly, it displaces the water above it, launching a series of rolling waves that will become a tsunami. About 80 percent of tsunamis are generated within the Pacific Ocean's "Ring of Fire," but they can occur in any large water body, including lakes.

Earthquakes can also dislodge rocks and sediment, causing landslides that similarly suddenly displace large amounts of water and generate what are known as megatsunamis. In 1958, a 7.8-magnitude earthquake in Lituya Bay, Alaska, caused a landslide that generated a 1,720-foot-high (524 meters) wave, the tallest ever recorded.

Tsunamis are usually made up of a series of waves known as a wave train. Each individual wave will have a very long wavelength. While a typical wind-generated wave might have a wavelength of about 165 feet (50 meters) and a height of about 7 feet (2 meters), a tsunami moving across the deep ocean may have a wavelength of 125 miles (200 kilometers) and a height of about 12 inches (30 centimeters). Hence, tsunamis typically go unnoticed at sea.

Because they have very long wavelengths, tsunamis tend not to lose much energy as they travel. They can move at speeds of up to 500 mph (800 km/h)—fast enough to cross the Pacific

Below *A tsunami flows over a sea wall near Miyako City, Japan, following the 9.1-magnitude Tohoku earthquake on March 11, 2011. The tsunami, which reached a height of about 36 feet (11 meters), was triggered by the rupture of a fault associated with the Japan Trench—which separates the Eurasian Plate from the subducting Pacific Plate—that saw part of the fault slip some 165 feet (50 meters).*

Ocean in less than a day—but because their wavelengths are so long, it might take as long as 20 or 30 minutes for the wave oscillation to complete a cycle from trough to trough.

As with normal ocean waves, when tsunamis reach shallower water, they begin to slow down and their height increases, but only the largest waves crest. Instead, they initially resemble a rapidly rising tide. However, they may still take several minutes to reach their full height, which can be in the tens of feet following large seismic events.

If the first part of the tsunami to arrive at the coast is a trough, a phenomenon known as drawback takes place. The edge of the sea recedes dramatically, sometimes by hundreds of feet, exposing areas that would normally be submerged for several minutes.

Both of these processes can cause considerable damage—the high-speed wall of water that accompanies a crest smashes into the coast and then the drawback pulls all of that water, now carrying a large amount of debris, back off the land. The damage can be extremely widespread, affecting entire ocean basins.

The tsunami that took place on December 26, 2004 in the Indian Ocean was one of the deadliest natural disasters in human history, with at least 230,000 people killed or missing in 14 countries. It was generated by a 9.1-magnitude earthquake that occurred 100 feet (30 meters) beneath the ocean floor off the Indonesian island of Sumatra and featured waves with heights of up to 30 feet (9 meters).

Tsunamis are very difficult to predict, even when details of the magnitude and location of an earthquake are known. Currently, automated systems that use pressure sensors attached to buoys offer the best protection, potentially providing warnings of an impending tsunami following an earthquake, enabling people to seek higher ground.

Below *The devastated southwest suburbs of Banda Aceh on the coast of Sumatra, Indonesia, lie underwater following the tsunami that struck on Boxing Day 2004. Banda Aceh, the closest major city to the epicenter of the 9.1-magnitude earthquake that triggered the tsunami, suffered severe casualties, with about 167,000 people losing their lives.*

// Coral reefs

Among the most productive and diverse ecosystems on Earth, coral reefs are also important physical structures that protect coastlines from damaging storms and erosion.

A coral reef growing in shallow water in the Solomon Islands. Because they rely on symbiotic algae for much of their food, most corals are found in warm, clear, shallow water.

Coral reefs are composed of the skeletons of living and dead colonial marine invertebrates. Corals belong to the phylum Cnidaria, which also contains jellyfish and sea anemones. Individual coral polyps are usually about 0.6 inches (1.5 centimeters) in diameter, but can be the size of a pinhead or as much as 12 inches (30 centimeters) across. They extract calcium carbonate from the surrounding seawater and use it build a hard, durable exoskeleton, each new polyp growing on an unoccupied part of the sea floor or, more often, on the skeletons of its predecessors.

There are four main categories of coral reef: fringing reefs, barrier reefs, platform reefs, and atolls. Fringing reefs, also called shore reefs, are the most common. They are generally found near coastlines, where they are either directly attached to the shore or separated from it by a shallow channel or lagoon. Barrier reefs form in open water and are separated from the coast by deeper, wider channels or lagoons. They take much longer to form than fringing reefs and are consequently much rarer. Platform reefs, also known as patch, bank, or table reefs, can form anywhere that the seabed rises close enough to the ocean's surface for warm-water reef-building corals to grow. In some cases they can be thousands of miles from land. Atolls are rings of coral around a central lagoon, formed when the island around which a fringing reef has grown erodes away and sinks back into the ocean. They can take as long as 30 million years to form.

Although reef-building corals grow all over the world at a wide variety of depths, the largest reefs are found in the warm (optimally 79–81°F/26–27°C), clear, shallow (less than 165 feet/50 meters deep), relatively nutrient-poor waters of the tropics and subtropics, between about 30°N and 30°S. The individual polyps on these reefs host microscopic single-celled algae called zooxanthellae

Above *A fringing coral reef growing around a small island in the Whitsunday group, Queensland. Growing out from the shore of an island, fringing reefs are the most common type of coral reef.*

within their tissues and these algae need sunlight to photosynthesize. The zooxanthellae produce carbohydrates that the coral polyps use for food, as well as oxygen, while the coral provides protection and the carbon dioxide that the algae need for photosynthesis. The algae provide as much as 90 percent of a polyp's energy needs; polyps catch the remainder with their tentacles, which bear stinging cells called nematocysts. It is the zooxanthellae that give most corals their colour.

Coral reefs are among the world's most diverse ecosystems, hosting countless individual organisms from thousands of different species. They provide food and shelter for about a quarter of all known marine species and are the primary habitat for more than 4,000 species of fish and 700 species of coral.

The combined area of the world's coral reefs is around 110,000 square miles (285,000 square kilometers), which amounts to less than 1 percent of the ocean floor. About 92 percent of that area is found in the Indo-Pacific region.

The world's largest reef system, the Great Barrier Reef, is located off the coast of Queensland, Australia. It is made up of more than 2,900 individual reefs and stretches for more than 1,600 miles (2,600 kilometers) over an area of almost 135,000 square miles (350,000 square kilometers).

Coral reefs are economically important. It has been estimated that, globally, reefs generate between US$30 billion and US$375 billion a year, mostly through fishing and tourism. They also protect coastlines from storms and erosion by absorbing wave energy, in some cases reducing it by 97 percent. More than half a billion people are thought to depend on reefs for their food, income and protection.

Above *As well as providing a highly productive, structurally complex habitat for a staggering diversity of marine organisms, many coral reefs provide protection for coastal areas.*

// Fjords

During the last ice age, massive glaciers carved deep, U-shaped valleys in the bedrock. When the glaciers melted, the sea level rose and water filled these coastal valleys, creating what we call fjords.

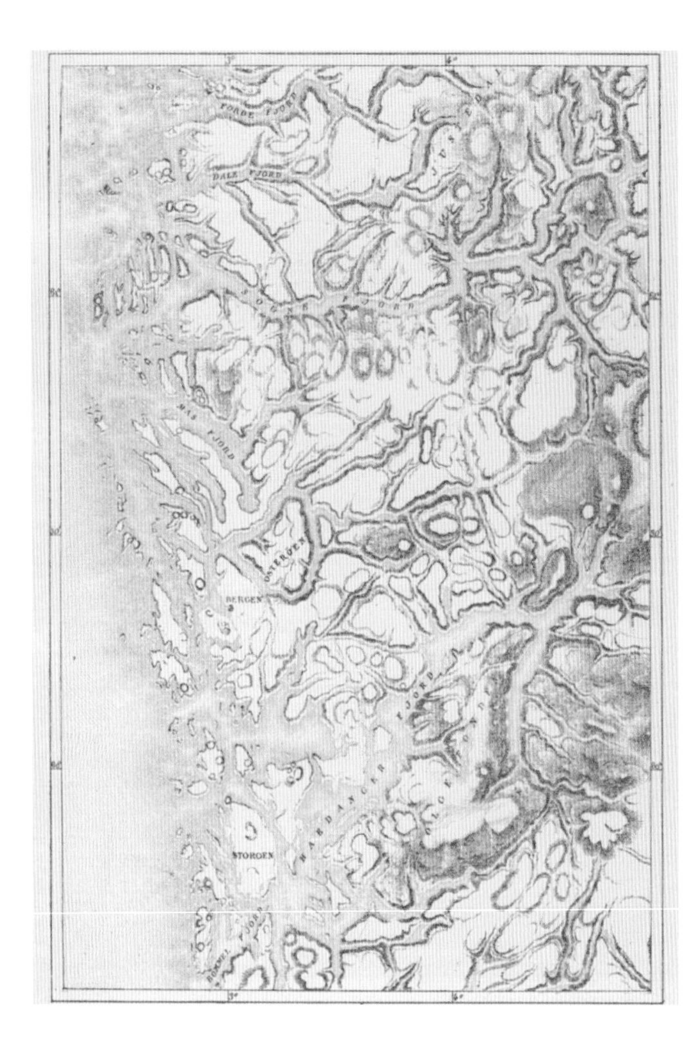

Above *This nineteenth-century map shows the extensive network of fjords in southern Norway.*

Many fjords are extremely deep, usually deeper than the adjacent sea. Their deepest point is typically at the inland end, where the glacial force was at its strongest. The world's deepest fjord is the Skelton Inlet on Antarctica's Ross Ice Shelf, which drops to 6,342 feet (1,933 meters) below sea level.

The mouth of a fjord is often its shallowest point, due to the deposition there of a moraine comprising gravel and sand picked up by the glacier during its slow journey to the coast. This shallow threshold means that the waters in a fjord are often calmer than in the open ocean—one reason why they make popular harbours for ships.

Fjords were created in the greatest abundance in regions where the prevailing westerly marine winds were forced upwards by mountains, causing them to dump the moisture that they were carrying as snow, feeding the glaciers flowing out of the mountains. Hence we now see the highest numbers of fjords in areas such as the west coast of Norway, the northwest coast of North America, the southwest coast of New Zealand and the southwest coast of South America.

Norway boasts almost 1,200 fjords, which account for 92 percent of the country's 18,000-mile (29,000-kilometer) coastline. However, the world's longest fjord is located in Scoresby Sund, a tree-like system of fjords that covers an area of about 15,000 square miles (38,000 square kilometers) in eastern Greenland. The longest branch in the system has a length of around 220 miles (350 kilometers).

Below *Geirangerfjord, Norway. A 9-mile (15-kilometer) branch of the 16-mile (26-kilometer) Sunnylvsfjorden, which is itself a branch of the 68-mile (110-kilometer) Storfjorden, the fjord is up to 853 feet (260 meters) deep and is surrounded by steep mountains.*

SKERRIES, REEFS AND RIAS

Skerries

Fjords are often accompanied by small rocky islands known as skerries. Remnants of the bedrock left behind as smaller glaciers gouged out a complex array of criss-crossing glacial valleys, skerries are typically found at the mouth of a fjord. Just off the coast of Norway is a remarkable collection of thousands of skerries, separated from the coast by a 1,000-mile (1,600-kilometer) channel made up of a series of deep fjords.

Skerries off the island of Rødøya, Norway.

Coral reefs

In the year 2000, scientists discovered extensive cold-water coral reefs living in the depths of Norway's fjords. Similar reefs have since been found in fjords in New Zealand. It is thought that the reefs may be at least partly responsible for the high productivity of Norway's fishing grounds.

Deep water coral in Trondheim Fjord, Norway.

Rias

Like fjords, rias are valleys that have been drowned by the rising sea. Unlike fjords, however, these valleys were carved from the rock by rivers, not glaciers. Rias often have a dendritic, or tree-like, structure formed by the small tributaries that prevously flowed into the river.

Ria of San Vicente de la Barquera, Spain.

// Sea ice

Frozen seawater that floats on the ocean surface at the poles, sea ice plays a central role in the global climate. But as the Earth heats up, it is disappearing at an alarming rate.

Sea ice is extremely dynamic—growing, melting, shifting position, and changing shape under the influence of winds, currents and temperature fluctuations.

When seawater begins to freeze, it forms tiny crystals called frazil. If the sea is calm, these crystals form slushy areas known as grease ice and then thin, smooth sheets of ice collectively called nilas. Once nilas has formed, water freezes to its underside, a process called congelation growth, and the ice sheets slide over each other to form rafts of thicker ice. In rough seas, waves and wind compress the crystals to form slushy pancakes that may be several meters in diameter and up to 4 inches (10 centimeters) thick. These collide with one another, sometimes sliding over each other to form smooth rafts or otherwise forming an upturned ridge around their circumference. They may eventually freeze together to form what is known as consolidated pancake ice.

Sea ice that is attached to the coastline or the sea floor is known as fast ice; if it drifts with the wind and ocean currents it is known as pack or drift ice. Individual pieces of sea ice 65 feet (20 meters) or more across are termed floes. Pack ice tends to be thicker than fast ice because pieces regularly collide, causing the floes to deform. This can create pressure ridges, hummocks, and what is known as rafted ice, when one floe is forced onto another.

Narrow, linear, dynamic openings in the pack ice are known as leads; they can be anything from a few feet to

Below *A satellite image showing the Arctic sea ice's minimum extent (1.8 million square miles/4.6 million square kilometers) in September 2017.*

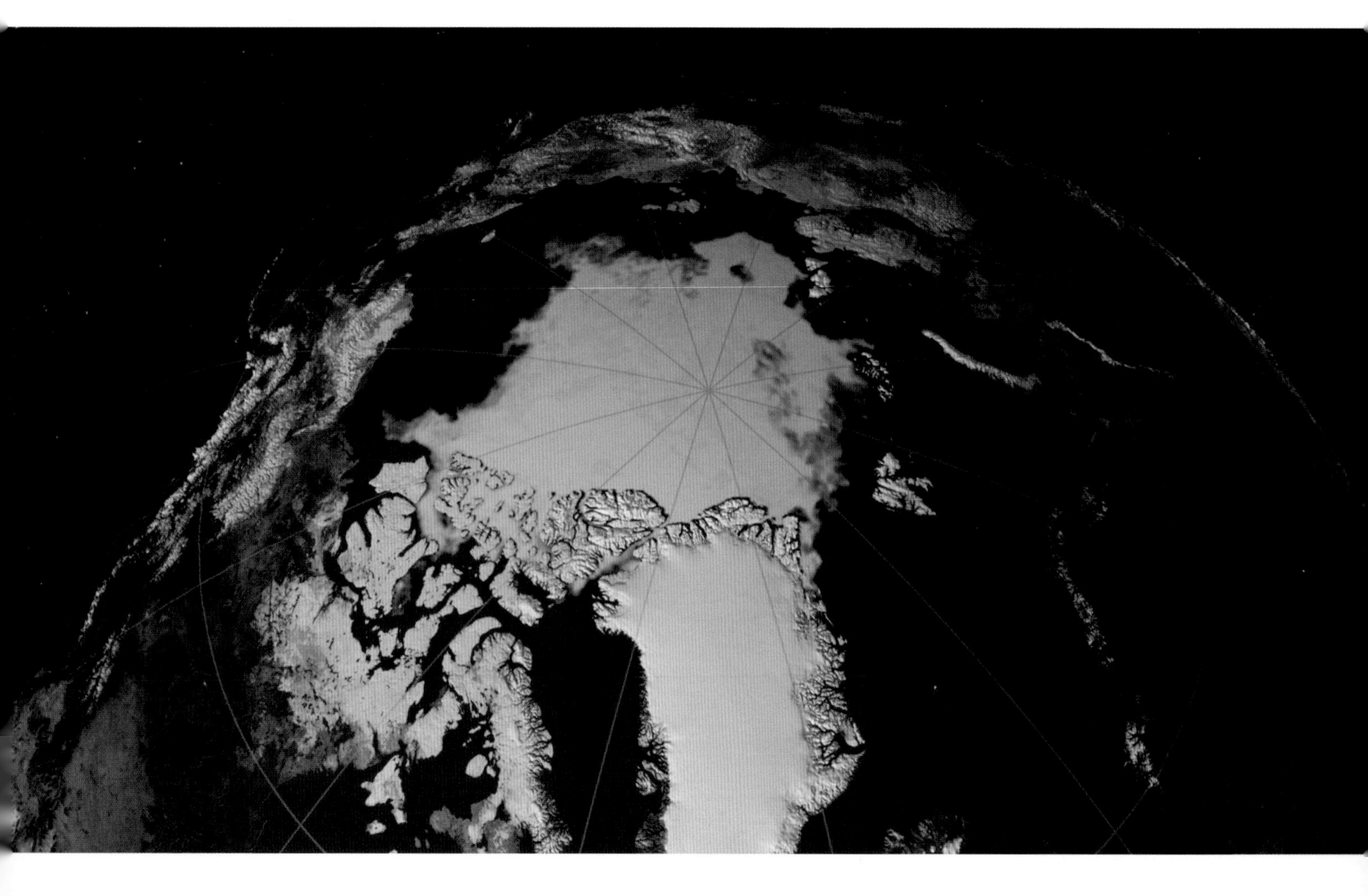

Above *Pancake ice off Svalbard, Norway. Jostling of the plates of ice has forced some to raft on top of the others and given them their characteristic raised rim.*

several miles in length and are continually forming and disappearing. Upwellings of warmer water and steady winds that blow ice away as quickly as it forms often create larger, more persistent openings called polynyas.

Ice shelves are the margins of continental ice sheets that extend into the ocean. They float in the ocean, but originate on the land.

If Arctic sea ice survives the summer melt season, it may last for many years. With each passing winter, it becomes stronger and thicker until it reaches a maximum thickness of 10–13 feet (3–4 meters); first-year sea ice is typically between 12 inches (30 centimeters) and 6.5 feet (2 meters) thick.

About 15 percent of the oceans (roughly 10 million square miles/25 million square kilometers) is covered with ice during at least part of the year. However, the area expands and contracts with the changing seasons.

Both the thickness and extent of Arctic summer sea ice have declined dramatically over the past 30 years, and global climate models suggest that the Arctic will be ice-free for at least part of the year before the end of the twenty-first century, possibly by mid-century. In Antarctica, sea ice melts almost completely each summer.

The melting and formation of sea ice does not influence sea level because it is already floating and hence displacing its own weight of water. However, the loss of sea ice has the potential to act as a powerful climate feedback (see page 184), accelerating global warming. Sea ice helps to keep the polar regions cool and dry. It acts as an insulating layer, reducing evaporation and preventing heat from escaping from the water. By limiting this heat exchange between the atmosphere and ocean, it has a profound effect on the global climate. In lowering the temperature of the adjacent air, sea ice increases the air-temperature difference between the tropics, subtropics, and polar regions, which affects atmospheric circulation patterns (see page 170).

THE AIR

We cannot see it, but without the air around us the Earth would be a very different place. The relatively thin blanket of gases that surrounds the planet is known as the atmosphere. Extending from the surface out to about 6,000 miles (10,000 kilometers) up, the atmosphere helps to support life on Earth in several ways: it forms an insulating layer that helps to keep us warm while also preventing extreme differences between day- and night-time temperatures; it shields us from most of the damaging ultraviolet radiation produced by the Sun; and, of course, it contains the oxygen that we breathe and the carbon dioxide that plants need to grow. It is also where our weather happens, and plays a vital role in the water and carbon cycles. The large-scale movement of air also plays a central role in the circulation of heat around the globe, which ultimately determines the climate in any given region. From down here on the Earth's surface, the atmosphere can seem so vast that there is little that we can do to affect it; but human activity has had a significant negative effect on the atmosphere. The release of carbon dioxide through the burning of fossil fuels is warming the planet and changing climate patterns around the world, while industrial pollution has damaged the protective ozone layer.

The "blanket" of gases that surrounds our planet helps to insulate us from temperature extremes while also sustaining life on Earth.

// The atmosphere

Held to the planet by gravity, the blanket of air known as the atmosphere protects and sustains life on Earth.

Left *The atmosphere eventually fades out about 62 miles (100 kilometers) above the surface of the Earth, a point known as the Kármán line, but about 98 percent of its mass is found in the lower 19 miles (30 kilometers).*

The atmosphere creates the pressure that allows water to exist in liquid form at the Earth's surface, absorbs damaging solar ultraviolet radiation, warms the surface by retaining heat, and reduces day–night temperature variation.

Today, by volume, dry air comprises 78.09 percent nitrogen, 20.95 percent oxygen, 0.93 percent argon, 0.04 percent carbon dioxide, and small amounts of other gases. The atmosphere around the primordial Earth was very different, consisting mostly of hydrogen, most likely supplemented by simple hydrides such as water vapor, methane and ammonia. Gases released during volcanic eruptions and the arrival of large asteroids—mainly nitrogen and carbon dioxide—came next; by about 3.4 billion years ago, nitrogen formed most of the atmosphere.

Things began to change significantly with the evolution of the first life forms and of photosynthesis, a by-product of which is oxygen. Most of this early oxygen was quickly consumed by the oxidation of minerals, particularly iron, but about 2.1 billion years ago, it started to build up in what is now known as the Great Oxygenation Event. By about 1.5 billion years ago, oxygen had reached a steady concentration of more than 15 percent and more complex life forms began to appear. Oxygen concentration has fluctuated since then, peaking at about 30 percent around 280 million years ago before dropping to today's 21 percent.

Water vapor, which makes up about 0.4 percent over the entire atmosphere, plays a significant role in regulating air temperature, absorbing both solar energy and thermal radiation from the Earth's surface. The atmosphere also contains numerous substances as aerosols, including organic and inorganic dust, pollen and spores, salt from sea spray, and volcanic ash. These aerosols play an important role in cloud formation (see page 148).

The total mass of the atmosphere is about 11.33×10^{18} pounds (5.14×10^{18} kilograms), around 2.9×10^{16} pounds (1.3×10^{16} kilograms) of which is water vapor. About 98 percent of that mass is found in the lower 19 miles (30 kilometers) of the atmosphere.

The atmosphere becomes thinner with increasing altitude, eventually fading out about 62 miles (100 kilometers) above the surface of the Earth, a point known as the Kármán line. The relative concentrations of different gases remain constant below a height of about 6.2 miles (10 kilometers) above the Earth's surface.

Atmospheric pressure, the total weight of the air above a particular point, varies with location and weather. The average atmospheric pressure at sea level is defined as 101,325 pascals, which is sometimes referred to as standard atmospheric pressure.

Close to the Earth's surface, air temperature is determined by three physical processes: radiation, conduction, and convection. A number of factors, such as wind, moisture, the physical characteristics of the surface below (whether land or water, for example), and the intensity of the sunlight, exert a further influence, determining the relative contributions of the three processes. Radiation is traditionally divided into shortwave radiation directly from

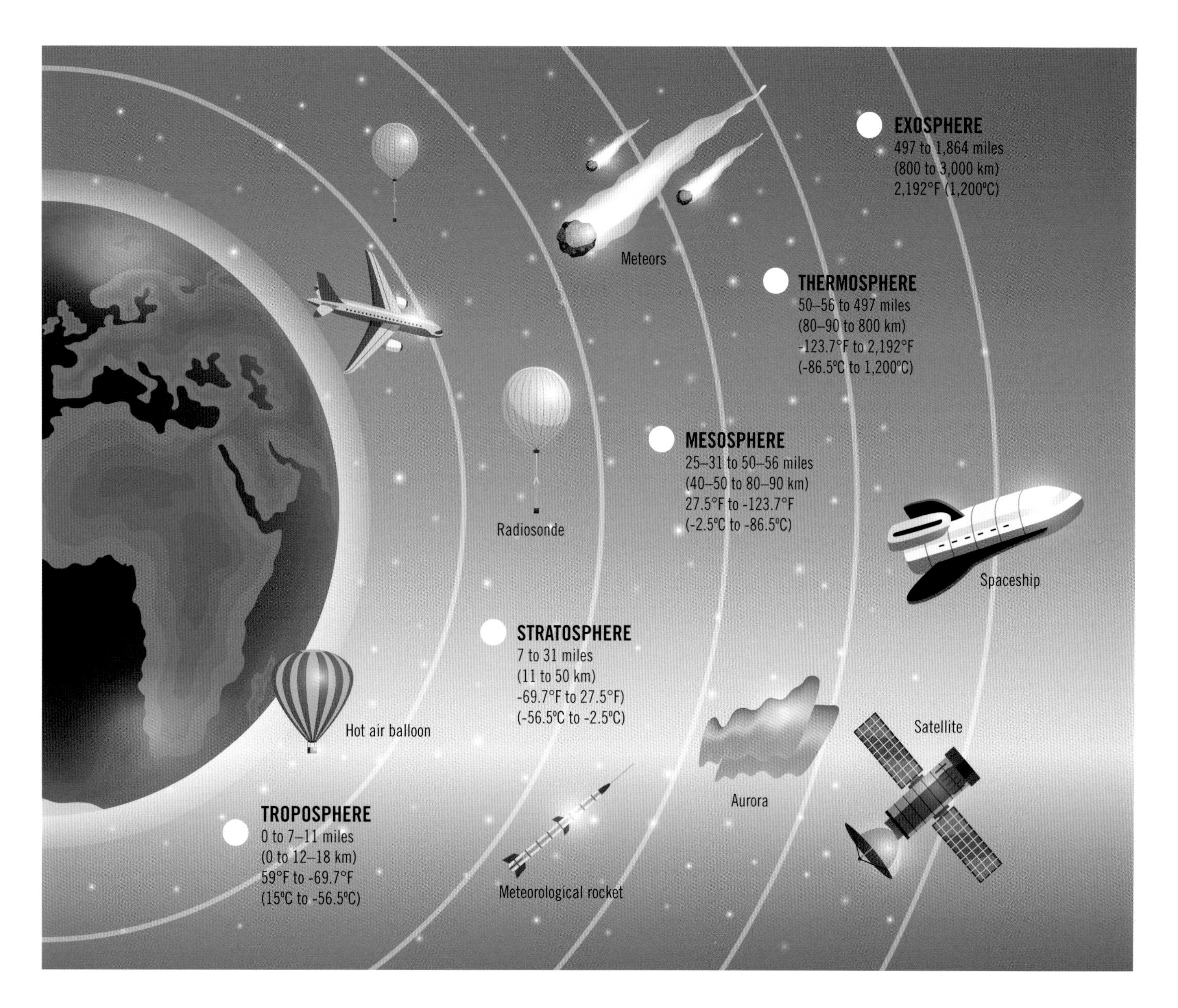

the Sun and longwave radiation emitted by the Earth's surface and the atmosphere itself.

If clouds are present, or the surface is covered by snow or ice, much of the incoming shortwave solar radiation is reflected back into space, but some warms the air as it is absorbed by atmospheric gases and some warms the land and ocean surfaces. The atmosphere and the Earth's surface then emit some of this energy as longwave radiation, which is absorbed by the "greenhouse gases" in the atmosphere—primarily water vapor and carbon dioxide—and re-emitted as more longwave radiation. This is why the Earth's surface cools faster on clear nights.

Light passing through the atmosphere is scattered by gas molecules. The shorter (blue) wavelengths are scattered more than longer (red) wavelengths, hence the sky appears blue. Sunsets appear red because the sunlight has to pass through more atmosphere, so most of the blue and green light is removed.

Human activity has changed the atmosphere through the release of pollutants, particularly since the Industrial Revolution. The emission of the greenhouse gases carbon dioxide and methane are causing the Earth's atmosphere to heat up, while the release of chlorofluorocarbons or CFCs have caused significant thinning of the ozone layer.

Above *The Earth's atmosphere is stratified. Scientists divide it into five layers based on characteristics such as temperature and composition: (from lowest to highest) the troposphere, stratosphere, mesosphere, thermosphere, and exosphere. The relationship between temperature and altitude is complicated, rising, falling, or staying constant with increasing altitude, depending on the distance from the surface; the boundaries of the layers correspond to temperature maximums and minimums.*

// The troposphere

The lowest layer of the Earth's atmosphere, the troposphere is a moist, turbulent region in which most of the world's weather systems reside.

The troposphere comprises about three quarters of the atmosphere's total mass and contains 99 percent of its water vapor. It is separated from the next major layer of the atmosphere, the stratosphere, by a thin layer of air called the tropopause. This is what is known as an inversion layer. Below it, the air temperature decreases with height; above it, the temperature increases with height. Within the tropopause, the temperature remains roughly constant at between -49°F (-45°C) and-112°F (-80°C) depending on the latitude. There is little mixing between the layers above and below. The jet streams (see page 174) are found in or just below the tropopause.

The troposphere is thinnest at the poles (about 4 miles/6 kilometers high) and thickest at the equator (about 11 miles/18 kilometers high). The height also depends on the season: it is lower during colder months. In both cases, this is due to there being less convection when the air and land are colder.

Within the troposphere, temperature, pressure and the amount of water vapor in the air all decrease rapidly with increasing altitude. The rate at which the temperature decreases is known as the environmental lapse rate and amounts to about 11°F (6°C) for every 0.6-mile (1-kilometer) increase in height. The concentration of water vapor in the troposphere also varies with latitude—it is highest over the tropics, where it can approach 4 percent, and decreases to a minimum over the poles.

The uneven solar heating of different parts of the

The space shuttle Endeavour *prepares to dock with the International Space Station over the South Pacific Ocean off the coast of southern Chile. The orange layer seen here is the troposphere.*

troposphere (greater at the equator than at the poles) drives the large-scale wind patterns that move heat and moisture around the world (see page 170).

When sunlight enters the atmosphere, some of it is immediately reflected back into space, mostly by clouds and snow and ice. The remainder reaches the Earth's surface, where it is absorbed and then re-emitted back into the troposphere as longwave radiation. Greenhouse gases in the troposphere, including carbon dioxide, methane, and water vapor, absorb this energy and emit much of it back towards the Earth (see page 182).

The lowest part of the troposphere is known as the planetary boundary layer. It varies in height from a few hundred meters to about 1.2 miles (2 kilometers), depending on the topography and the time of day. This is the level where friction with the Earth's surface influences the movement of air, and where air is warmest, heated from below by latent heat (energy absorbed or released when a substance changes its physical state without changing its temperature), longwave radiation, and sensible heat (thermal energy whose transfer to or from a substance leads to a change of temperature). Convection, whereby warm air expands and rises, controls the way that gases behave in this layer, maintaining a vertical temperature gradient from warm at the surface to colder higher up.

Above *An anvil cloud is a type of cumulonimbus cloud formed when the rising air in the cloud's characteristic "tower" reaches the tropopause, where the strong temperature inversion stops further upward motion and the cloud's top flattens and spreads out into an anvil shape.*

// The stratosphere

The second layer of the atmosphere as you travel away from the Earth, the stratosphere is home to the protective ozone layer.

The stratosphere sits between the troposphere below and the mesosphere above, roughly 6–30 miles (10–50 kilometers) above the Earth's surface. Its upper boundary is known as the stratopause.

About 10 percent of the mass of the Earth's atmosphere is found in the stratosphere. The air at the top of the stratosphere is roughly 1,000 times thinner than the air at sea level.

Water vapor is extremely rare in the stratosphere. All air that makes its way into the stratosphere has to pass through the tropopause, where the extremely low temperatures essentially freeze out all water. Occasionally, however, the tops of tall cumulonimbus clouds ascend through the tropopause and into the stratosphere. In winter, polar stratospheric clouds (also known as nacreous clouds) also sometimes appear in the lower stratosphere over the poles. They form at altitudes of between 9 and 16 miles (15 and 25 kilometers) when temperatures at those heights drop below -108°F (-78°C). These clouds play a role in the creation of the holes in the ozone layer by facilitating certain ozone-destroying chemical reactions (see page 136).

Unlike in the troposphere, temperature in the stratosphere decreases with increasing altitude. In the lower portion, near the tropopause, the temperature is a near-constant -58°F (-50°C) or so, rising to an average of about 14°F (-10°C) at the stratopause. This is due to the presence of the ozone layer.

Ozone is concentrated at about 9 miles (25 kilometers) above the Earth's surface. The continuous destructive and creative processes of ozone's formation heat what little air is present. The higher the ozone concentration, the higher the temperature.

Above *Polar stratospheric or nacreous clouds over Norway. Forming in winter in the lower stratosphere, such clouds may consist of supercooled droplets of water, nitric acid, and/or sulfuric acid, or ice crystals.*

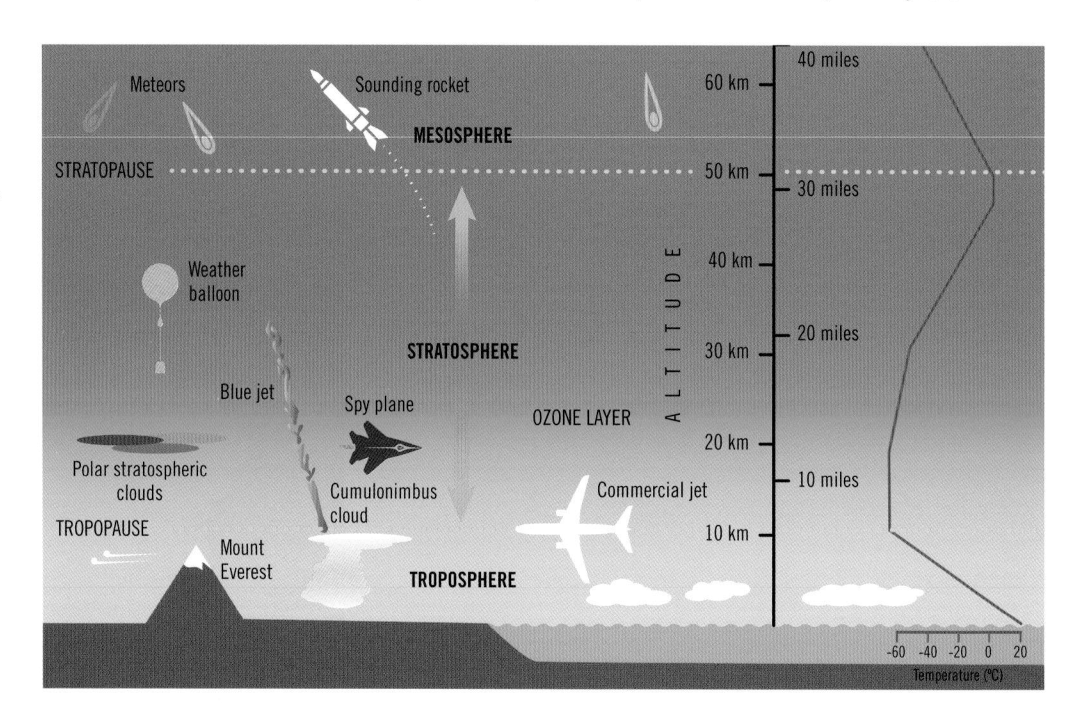

Above *The stratosphere stretches from about 6 to 31 miles (10 to 50 kilometers) above the Earth's surface. The air temperature within this layer decreases with increasing altitude.*

Above *The eruption of Mount Pinatubo in the Philippines in 1991 ejected almost 20 million tonnes (22 million tons) of sulfur dioxide into the stratosphere, causing global temperatures to drop by about 0.9°F (0.5°C).*

The temperature gradient and stratification within the stratosphere is extremely stable, with very little convection and mixing taking place. The exceptions are when large volcanic eruptions and the tops of supercell thunderstorms temporarily reach stratospheric heights.

Because there is little in the way of vertical convection, any material that makes its way into the stratosphere tends to stay there for a long time. This includes ozone-destroying CFCs and aerosols ejected in large volcanic eruptions, which can have a significant impact on climate.

The lack of vertical convection also means that turbulence is rare in the stratosphere. What little there is is caused by perturbations in the jet streams that blow around the Earth in the upper troposphere (see page 174). The stratosphere does have its own complex wind systems nonetheless, and winds the can reach speeds as high as 140 mph (220 km/h); however, violent storms do not occur in the stratosphere.

// The ozone layer

Within the stratosphere is a layer of gas that is particularly important to life on Earth: the ozone layer or ozonosphere.

Ozone is a molecule of oxygen that contains three oxygen atoms, rather than the more common two. At ground level, ozone is an air pollutant (it is one of the main ingredients in photochemical smog), but in the stratosphere it acts as the Earth's sunscreen, shielding it against incoming ultraviolet (UV) radiation from the Sun, which can damage animal and plant cells. The ozone layer absorbs about 98 percent of the incoming UV light.

Stratospheric ozone is created when UV light splits ordinary oxygen molecules (O_2) into individual oxygen atoms (atomic oxygen) that then combine with other intact O_2 molecules to form ozone (O_3), releasing heat as they do so. Ozone molecules are constantly being destroyed and reformed naturally, both processes triggered by UV light.

Ozone is very rare in the atmosphere; within the ozone layer itself, it occurs at less than 10 parts per million, while the concentration in the Earth's atmosphere as a whole is about 0.3 parts per million. About 90 percent of the ozone in the atmosphere is found within the stratosphere.

The thickness of the ozone layer varies both seasonally and geographically, but it typically occupies the lower portion of the stratosphere, from about 9 to 22 miles (15 to 35 kilometers) above the Earth's surface. Most ozone is produced above the tropics and then carried towards the poles by stratospheric winds. In general, it tends to be thinner over the equator and thicker over the poles. This variation is due to a combination of atmospheric circulation patterns and the intensity of solar radiation.

Below *The ozone layer blocks out all harmful UV-C radiation, while letting in some of the less harmful UV-B and all UV-A radiation.*

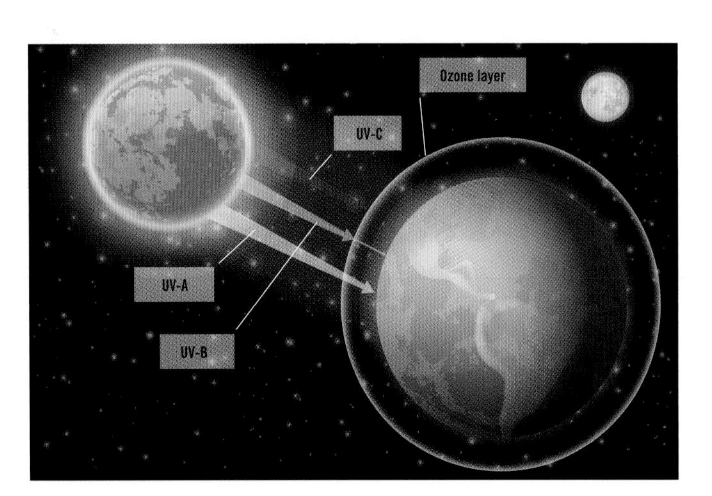

Below *This cross section of the Earth's atmosphere, as measured by NASA's Suomi NPP satellite, shows the ozone layer in red and orange.*

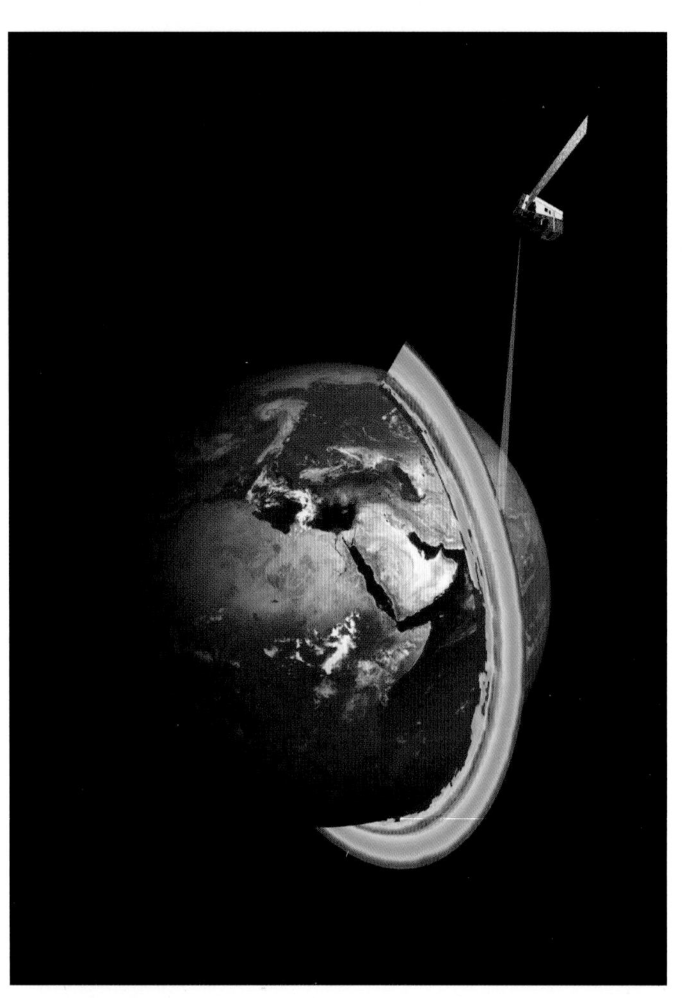

Ozone 'holes'

In 1976, atmospheric scientists discovered that the ozone layer was becoming thinner; since then, global ozone levels have dropped by about 4 percent. Among the most significant causes of this thinning were industrial chemicals known as chlorofluorocarbons, or CFCs. CFCs drift up into the upper atmosphere and are eventually broken down by UV light, causing them to release chlorine atoms that then react with ozone, stripping away one of the oxygen atoms. The chlorine eventually reacts with hydrogen-containing compounds to form HCl (hydrochloric acid), which is water soluble. It thus precipitates out of the atmosphere in water droplets or ice crystals. A single chlorine atom can destroy more than 100,000 ozone molecules before it is removed from the stratosphere.

The problem is particularly acute over the poles, where "holes" open up each spring (technically, they are not holes, but rather regions where the ozone layer is much thinner). CFCs break down more rapidly when ice crystals are present, which is why the holes are situated over the poles; the hole above the South Pole is particularly large because icy clouds tend to be more common over Antarctica.

Below *The ozone hole over Antarctica. In 1979, Antarctic ozone levels dropped below 200 Dobson Units for the first time on record (Dobson Units are a measure of the amount of a trace gas in a vertical column through the Earth's atmosphere). During the 2008 Southern Hemisphere spring, they reached the lowest level ever recorded at just 100 DU, but with the ban on CFCs now in place, levels are slowly returning to normal. However, a full recovery isn't expected to occur until 2040.*

Although the global ban on CFC production negotiated under the 1987 Montreal Protocol has seen the ozone layer begin to bounce back, CFCs potentially have atmospheric lifetimes of more than a century, so it may take until the middle of the twenty-first century for the ozone layer to return to 1980 levels.

OZONE AND LIFE ON EARTH

The ozone layer is considered by some scientists to have been crucial to the development of life on Earth. About two billion years ago, a rise in atmospheric oxygen concentrations led to a build-up of ozone in the Earth's atmosphere, screening out UV-B radiation, which at that time was at lethal levels, and thus facilitating the movement of life from the oceans to the land.

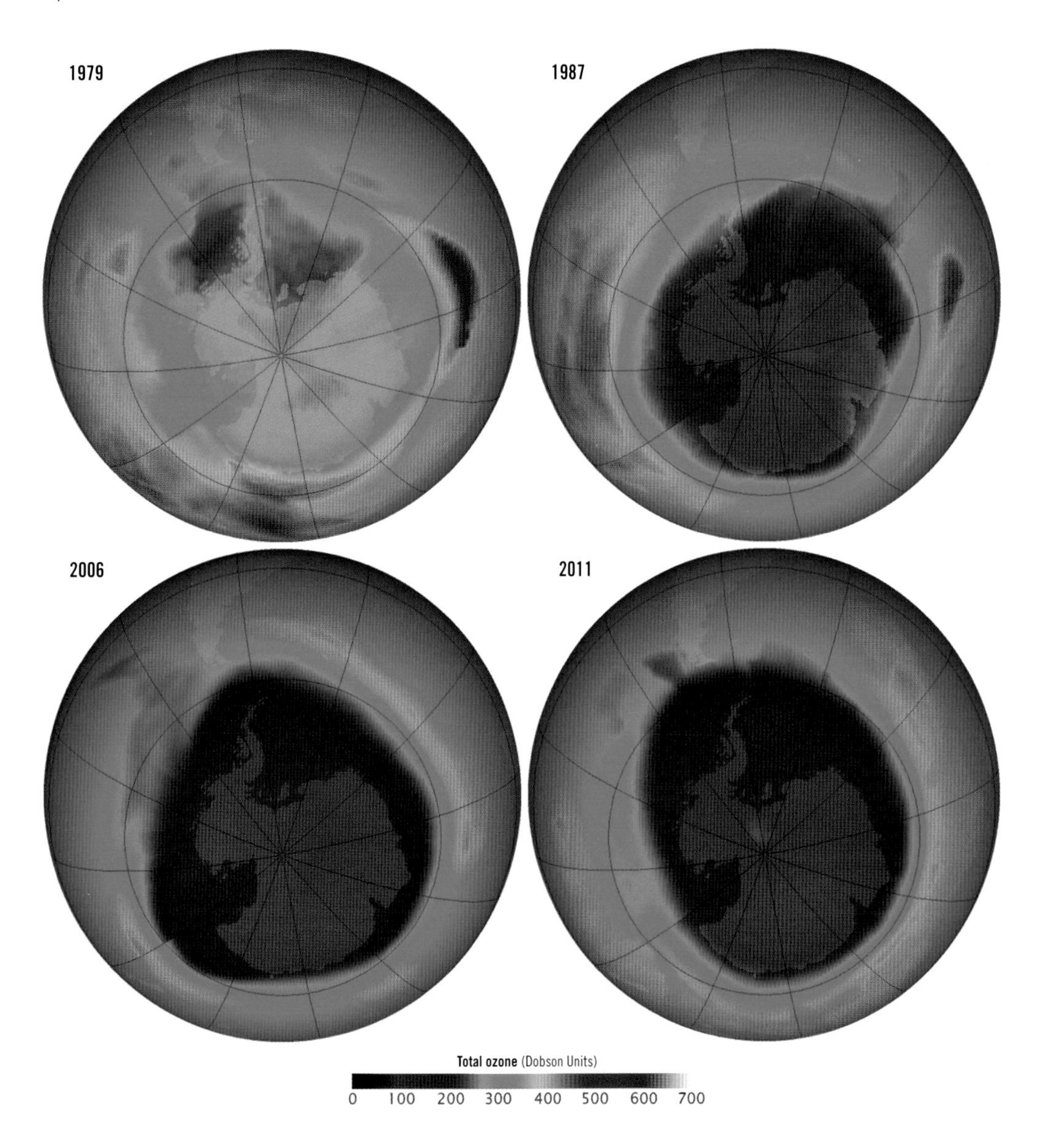

// The mesosphere, thermosphere, and exosphere

In its outer three layers, the atmosphere becomes progressively thinner until it eventually merges with outer space.

Geminid meteors leave streaks of light in the night sky as they burn up in the mesosphere.

The mesosphere

Sitting above the stratosphere, the mesosphere extends from about 31–53 miles (50 to 85 kilometers) above the Earth's surface. Its upper and lower boundaries vary with latitude and season: they are higher in winter and over the tropics, lower in summer and over the poles. At the base of the mesosphere, air pressure is about one thousandth that at sea level; at its top, it is about one millionth—effectively a vacuum.

As in the troposphere, temperature decreases with increasing height throughout the mesosphere. The coldest temperatures found within the Earth's atmosphere—as low as -130°F (-90°C)—are in the mesopause, the mesosphere's upper boundary. What little atmosphere exists in the mesosphere is influenced by winds, waves, and tides that carry energy upward from the troposphere and stratosphere.

The mesosphere is where shooting stars occur—most of the millions of meteors that enter the atmosphere each year are vaporized within the mesosphere due to collisions with gas molecules and it consequently has a relatively high concentration of iron and other metal atoms. The material left behind is known as meteoric smoke.

Occasionally, noctilucent clouds—"night-shining" clouds that are only visible at twilight and are actually composed of frozen water vapor—form in the mesosphere near the poles, probably around meteoric smoke. Electrical discharges similar to lightning, known as sprites and elves, also occasionally appear in the mesosphere, some distance above thunderclouds in the troposphere.

The mesosphere is difficult to study because it is too high for weather balloons and powered aircraft to reach and too low for satellite orbits.

The thermosphere

The thickest layer of the atmosphere, the thermosphere sits above the mesosphere. It extends from about 56 miles (90 kilometers) to 310–620 miles (500–1,000 kilometers) above the Earth. The thermosphere's upper limit, the thermopause, varies in height depending on the

Above *Noctilucent clouds over Stockholm, Sweden. Located in the mesosphere at altitudes of around 47–53 miles (76–85 kilometers), noctilucent clouds consist of ice crystals and are only visible during twilight, generally only during summer.*

level of solar activity: when the Sun is particularly active, the thermosphere grows hotter and expands, causing the thermopause to move upwards.

Most of the highly energetic solar radiation (X-rays and extreme ultraviolet radiation) that strikes the Earth is absorbed by the thermosphere. Consequently, temperatures increase with altitude and are highly dependent on solar activity, ranging from about 437°F (225°C) during periods of low sunspot activity to 3,630°F (2,000°C) or higher when the Sun is more active. In the lower thermosphere, below an altitude of 125–185 miles (200–300 kilometers), temperatures increase rapidly with height, before levelling off and holding relatively steady. The impact of solar radiation is such that the daytime temperature is usually about 360°F (200°C) hotter that the temperature at night. However, because the air is so thin in the thermosphere, temperature in the usual sense is not very meaningful as collisions between gas molecules are so rare that heat transfer is almost non-existent.

As in the mesosphere, the thermosphere's rarefied atmosphere is moved around by winds, waves and tides that originate in the troposphere and stratosphere. Although there is still enough atmosphere in the thermosphere for gas molecules and atoms to collide with each other, it happens very infrequently and the gases become somewhat separated based on the types of chemical elements they contain. In the upper thermosphere, atomic oxygen, nitrogen, hydrogen, and helium are the main atmospheric components.

The exosphere

The exosphere is the uppermost layer of the Earth's atmosphere, where it gradually fades into the vacuum of space. The bottom of the exosphere, sometimes called the exobase, ranges from about 310 to 620 miles (500 to 1,000 kilometers) in altitude, depending on solar activity. The atmospheric temperature is a near-constant 2,237°F (1,225°C) above this altitude.

There is disagreement over whether the exosphere forms part of the Earth's atmosphere; some scientists consider it to be just part of space. Because it gradually fades into outer space, the exosphere has no clear upper boundary, but is generally considered to reach 6,200 miles (10,000 kilometers) above the Earth's surface.

Air in the exosphere is so thin that collisions between gas molecules and atoms (mostly hydrogen, helium, and some carbon dioxide, nitrogen, and oxygen) are very rare. Instead, they move around the Earth along curved paths, eventually either falling back into the lower atmosphere under the influence of gravity or, if their speeds are great enough, flying off into space.

The orbits of many satellites lie in the exosphere or below. The vestigial atmosphere in the exosphere is enough to create a tiny amount of drag on these satellites, causing them to slow and eventually fall out of orbit.

Below *Polar mesospheric clouds illuminated by the rising Sun, as seen from the International Space Station.*

// The homosphere, heterosphere, and ionosphere

The atmosphere can also be divided up according to the degree to which its gases are mixed and the impact of the Sun's rays on atoms of those gases.

The homosphere and heterosphere

In addition to the five layers discussed previously, the Earth's atmosphere can also be divided into two major zones based on how much its gases are mixed and hence how uniform it is. The lower of these two layers is called the homosphere. Here the atmosphere is dominated by turbulent mixing, which ensures that its composition is generally uniform, independent of height.

The upper limit of the homosphere, known as the turbopause, is found just above the mesopause, at an altitude of roughly 62 miles (100 kilometers). The region above this is called the heterosphere. Here, molecular diffusion dominates and the chemical composition of the atmosphere varies with altitude according to the atomic weights of the different gases present, with oxygen and nitrogen present only in the lower levels and only hydrogen present in the upper levels.

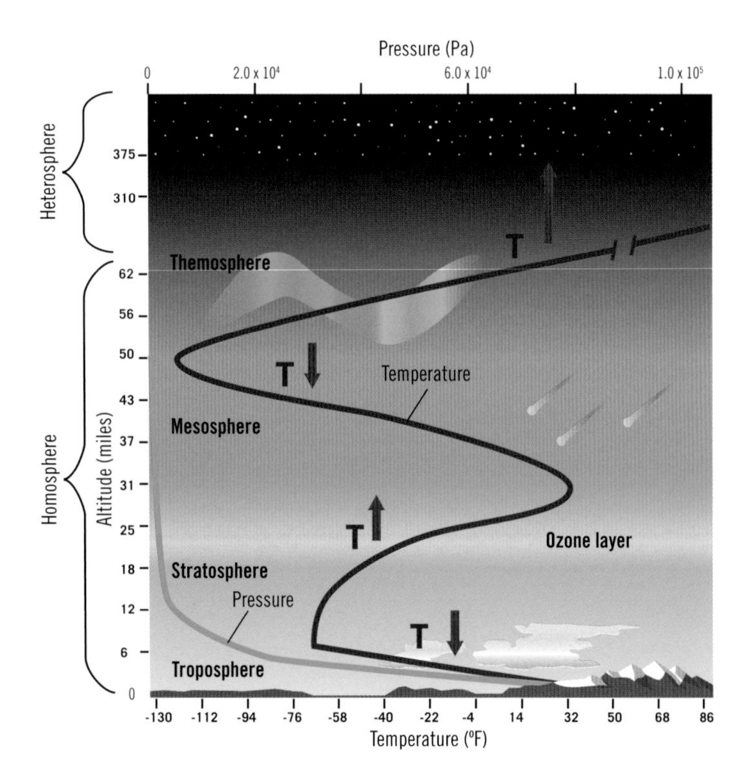

Above *The homosphere, the layer of the atmosphere dominated by turbulent mixing, stretches from the surface to about 62 miles (100 kilometers) up. Above this lies the heterosphere, where molecular diffusion regulates the atmosphere's chemical composition.*

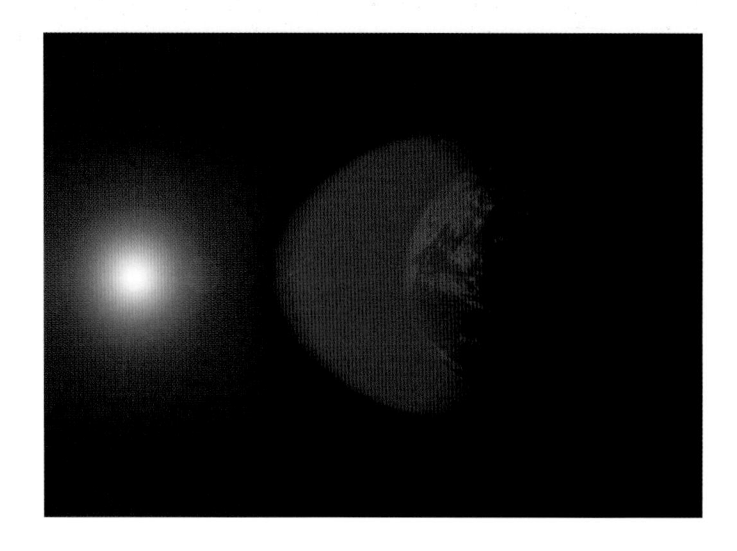

Above *The ionosphere swells out in response to the Sun's rays.*

The ionosphere

The ionosphere is located in the Earth's upper atmosphere, extending from the top half of the mesosphere through the thermosphere to the lower part of the exosphere—from 50 to 620 miles (80–1,000 kilometers) above the surface. Here extreme ultraviolet and X-ray solar radiation strips electrons from atoms and molecules in the atmosphere to form ions. The ionosphere has practical importance because it influences the way in which radio signals propagate and hence the distance they can travel.

The concentration of ions and free electrons depends on the amount of solar radiation striking the atmosphere, so the ionosphere is dense with charged particles during the day. During the night, ions recombine with displaced electrons and their concentration drops, causing entire layers of the ionosphere to appear and disappear over the daily cycle. There is also a seasonal effect, with ionization reaching a peak in summer and dropping to a low in winter. And finally, the density of ions fluctuates with the Sun's 11-year cycle of activity; solar flares and associated changes in the solar wind and geomagnetic storms increase the amount of ionization.

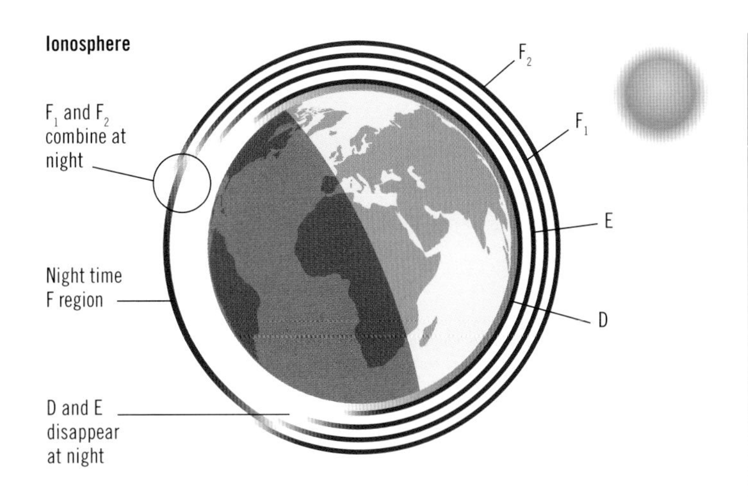

Above *The layers of the ionosphere during the day and night.*

Above *The ionosphere and aurora as seen from the Internation Space Station.*

The ionosphere itself can be divided into three layers: D, E and F. As in the other parts of the atmosphere, the depths of the layers vary with time of day, season and latitude.

The innermost layer, the D layer, extends from 37 to 56 miles (60–90 kilometers) above the Earth's surface. Due to high recombination rates, there are more neutral air molecules than ions in this layer. The D layer absorbs medium-frequency and lower-high-frequency radio waves, particularly at 10 MHz and below. This effect is higher during the day and can cause distant AM radio stations to disappear before reappearing with extended ranges at night.

The E layer is the middle layer, located 56–93 miles (90–150 kilometers) above the Earth's surface. Like the D layer, it weakens at night and its region of greatest strength rises higher. The E layer normally reflects radio waves with frequencies below about 10 MHz, but when the layer is particularly strong, it can reflect frequencies up to 50 MHz and higher.

The F layer extends from about 93 miles to more than 310 miles (150–500 kilometers) above the Earth's surface. Electron density peaks in this layer. At night, the F layer consists of a single layer (F2); however, during the day, a second, weaker layer (F1) often forms. Because the F2 layer is always present, it is mostly responsible for the refraction and reflection of radio waves.

The upper sections of the ionosphere overlap with the lower part of the magnetosphere, the region where charged particles interact strongly with the magnetic fields of the Earth and the Sun. It is here that the auroral displays take place (see page 180).

Left *During the day, VLF (very low frequency) radio waves bounce off the ionosphere's D layer. At night, the D layer disappears and they are reflected by the higher E and F layers, which means that they will travel much farther.*

// Climate

The atmospheric conditions that prevail in a given region determine its climate—the sort of weather it experiences—which shapes the world's animal and plant communities. Climate is distinct from weather. Weather is what we experience over the short term; climate is what a region experiences over the longer term. The difference is neatly summed up by the phrase: "Climate is what you expect, weather is what you get."

Climate is often described as the average weather, but it is much more than that, encompassing the average weather conditions as well as the extreme ranges, variability, and frequency of various weather-related phenomena, measured over a long period, from months to millions of years. When discussing modern climate, a period of 30 years is typically used—long enough to filter out year-to-year variation and anomalous phenomena, but short enough for longer climatic trends to be discerned. Among the meteorological variables commonly measured are temperature, humidity, atmospheric pressure, wind, and precipitation.

Below *A satellite image of global vegetation. Climate is the major determinant of the amount and type of vegetation in a given region.*

Climates generally change over relatively long time scales—thousands to hundreds of thousands of years—but can alter more rapidly as a result of variations in ocean circulation or volcanic activity. Some of these changes occur on regular cycles, driven by alterations in the Sun's output or the shape of the Earth's orbit, for example. However, since the Industrial Revolution, the global climate has been changing relatively rapidly, mainly due to the build-up in the atmosphere of greenhouse gases released by human activity (see page 182).

The climate of a given location is determined by several factors, including latitude, local topography and vegetation cover, altitude, the ratio of land to water, and the proximity to large water bodies and their temperature. Latitude is a

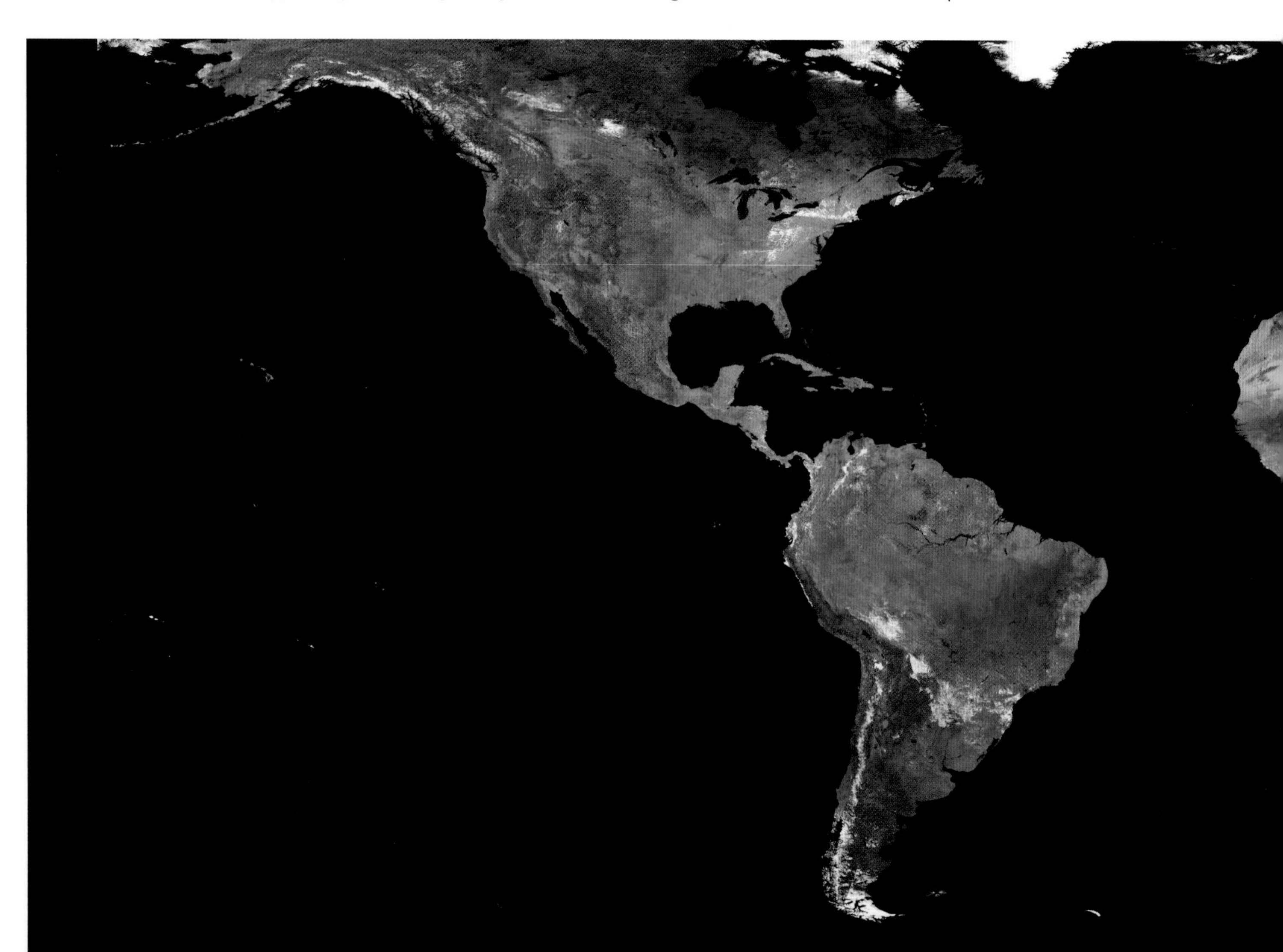

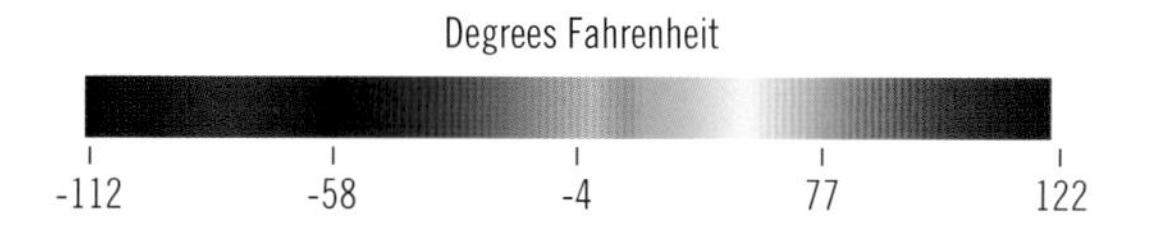

Above *Global temperature map. The effect of latitude on local temperatures is clear, as is the impact of the bands of high pressure that encircle the Earth at latitudes between about 15° and 30° north and south, within which the highest temperatures are observed.*

factor because of its impact on the angle at which the Sun's rays strike the Earth, which in turn affects the intensity of solar radiation and hence the local temperature. Winds and ocean currents also play a significant role in determining a region's climate as they move heat and moisture around, and are often the progenitors of storms.

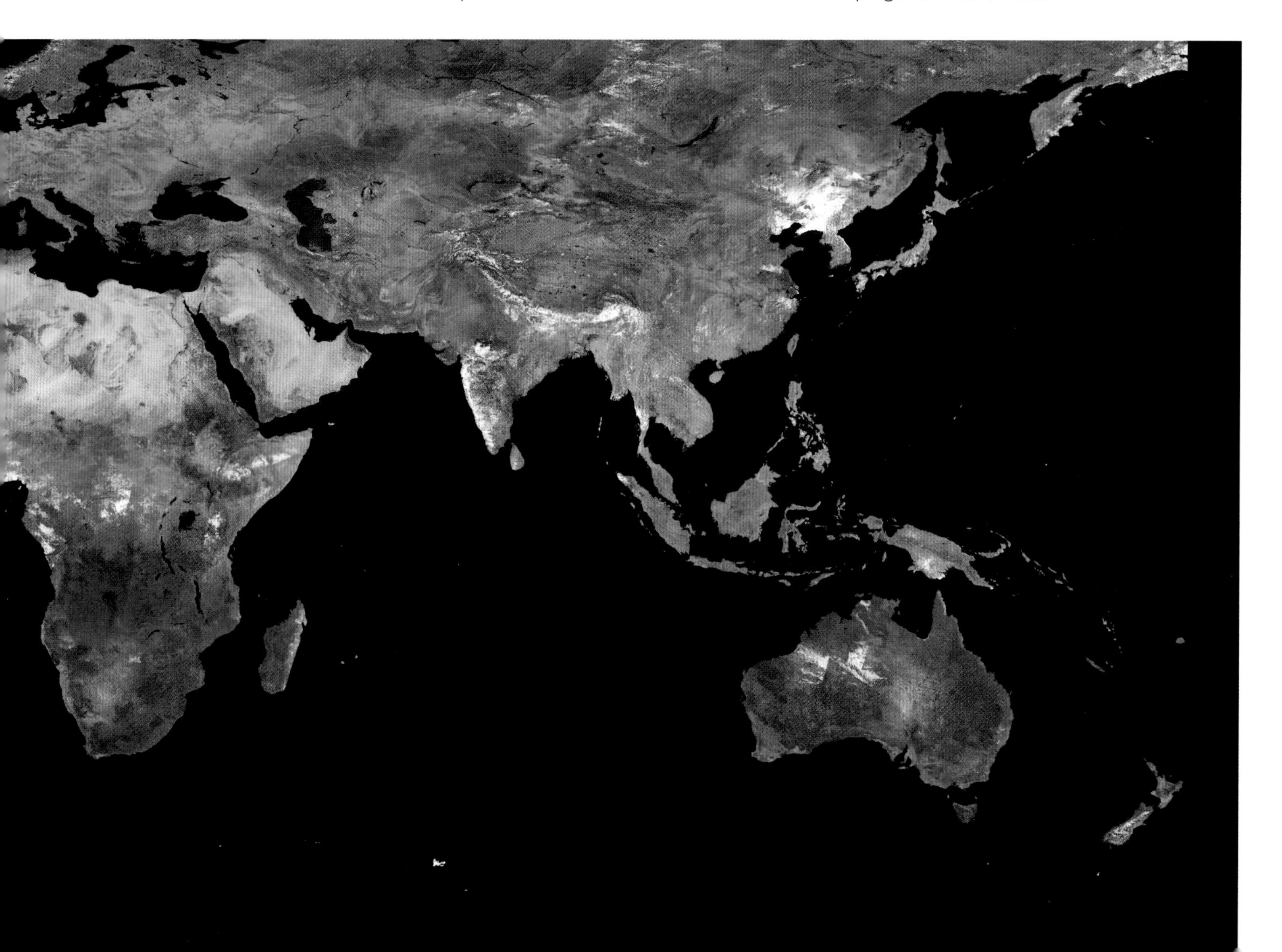

// Climatic zones

The Earth can be divided up into different zones based on the nature of the local climate.

The most commonly used classification for the world's climatic zones was developed by Wladimir Köppen (1846–1940), a Russian climatologist and amateur botanist of German descent. The Köppen system divides the world into five major regions based on a combination of vegetation type, precipitation, and temperature. Each of these five zones is given a capital letter, A to E, and is divided into subcategories based on the temperature, the amount of precipitation, and the season in which that precipitation occurs. So, for example, Af indicates a tropical rainforest climate and Dwa indicates a continental climate with a dry winter and a hot summer.

For the most part, the different climatic zones sit within particular bands of latitude and in similar positions on the continents. The exception is the continental zone, which is not found at higher latitudes in the Southern Hemisphere because there are no landmasses large enough to produce a continental climate.

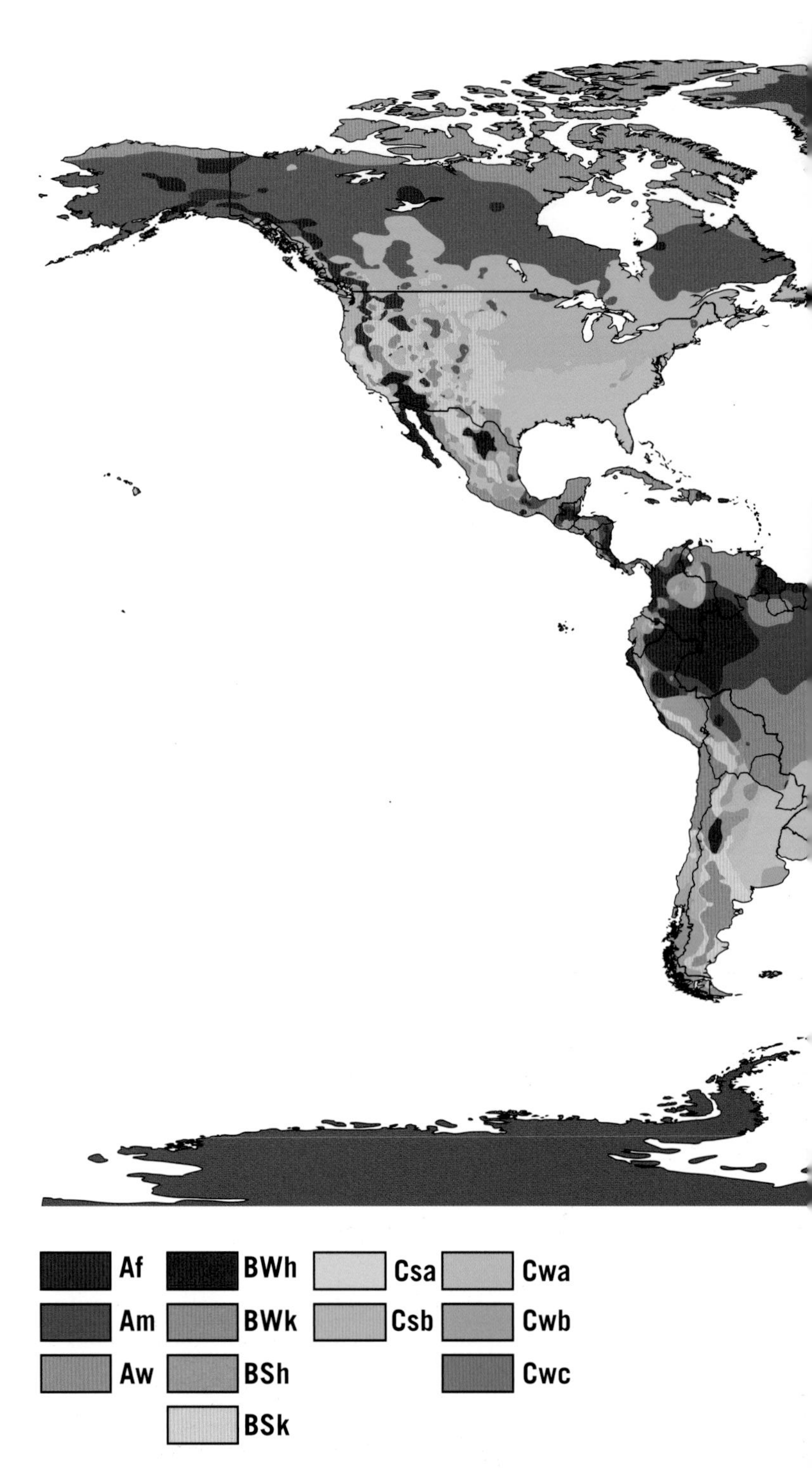

Tropical (A): Temperatures are high but not extreme, typically ranging from 25°C to 35°C (77–95°F), and show little in the way of seasonal variation; monthly average temperatures are 18°C (64°F) or higher. Day and night length also show little variation. Rainfall is usually high, as is humidity. Divided into rainforest, monsoon, and wet and dry savanna climates.

Arid (B): Annual precipitation is low. Divided into hot desert, cold desert, hot semi-arid/steppe, and cold semi-arid/steppe, as well as climates characterized by frequent fog.

Temperate (C): Generally a mild climate with warm to hot summers and cool winters. Experiences four distinct seasons. The average temperature during the coldest month is between 32°F and 64°F (0–18°C), and is above 50°F (10°C) during at least one month. Divided according to the precipitation pattern (dry winters or summers, or significant precipitation in all seasons) and the level of summer heat (hot, warm, or cool).

Continental (D): Summers are warm to hot and winters are cold. At least one month has an average temperature below 32°F (0°C) and at least one month has an average temperature above 50°F (10°C). As the names suggests, these climates usually occur in the interiors of continents. They are rare in the Southern Hemisphere due to the smaller land masses in the mid-latitudes. Divisions are as for Temperate, with the addition of a category for very cold winter climates.

Polar (E): Extremely cold; monthly average temperatures never exceed 50°F (10°C). Divided into tundra (average temperature of warmest month between 32 and 50°F (0–10°C) and ice cap (average temperatures never exceed 32°F/0°C).

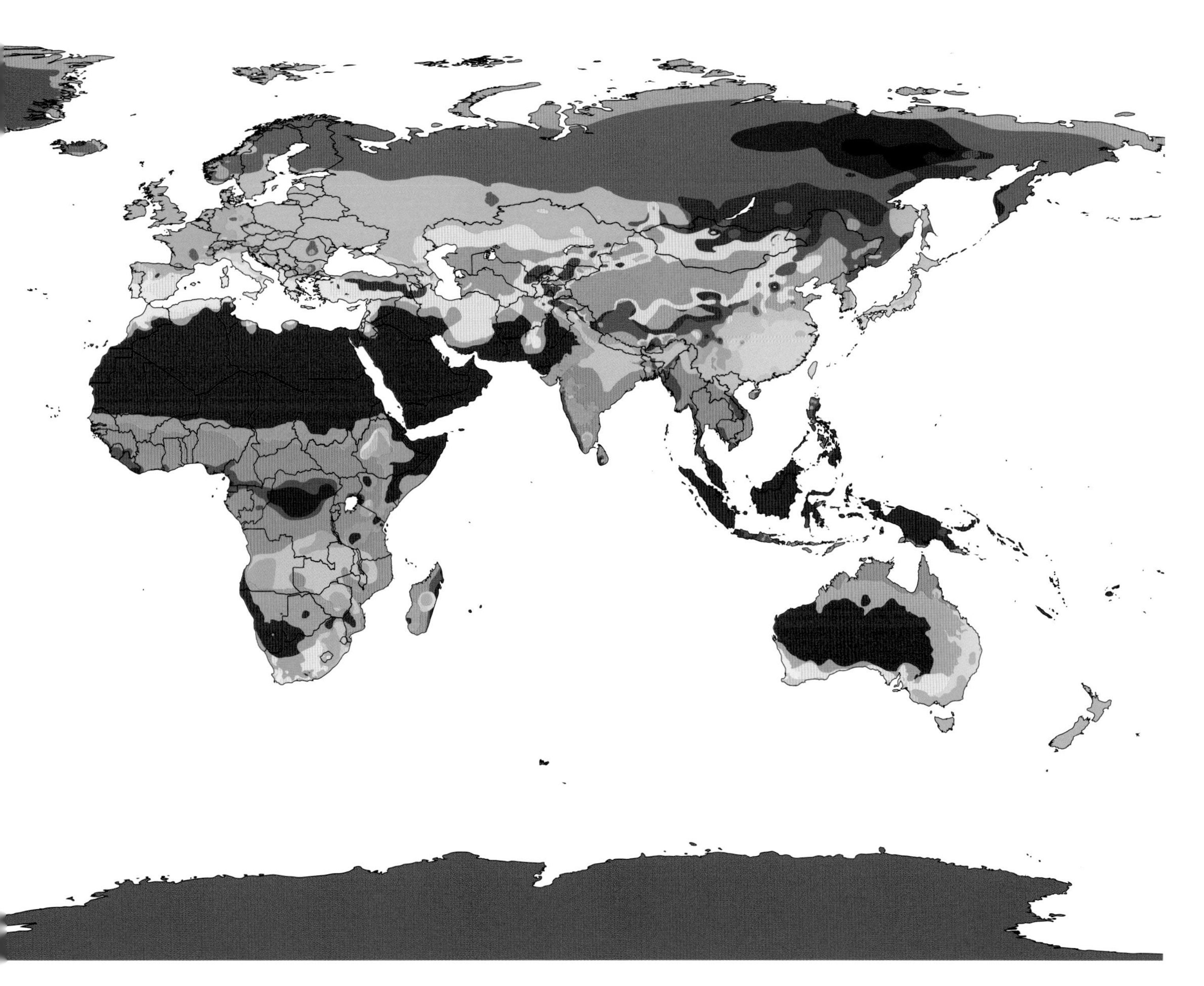

Above *World map showing the Köppen-Geiger climate classification, an update of the original Köppen classification produced by climatologist Rudolf Geiger in 1961.*

BIOMES

Scientists also divide the world into what are known as biomes—communities of plants and animals that share common characteristics as a result of the types of environment in which they live. So, for example, organisms that live in the desert biome will have certain similarities, regardless of which desert on which continent they are found in. Among the other biomes are tropical rainforest, grassland, savanna, and tundra.

// The water cycle

The constant movement of the Earth's water between the surface and the atmosphere is known as the water cycle and is driven by the Sun's energy.

At its simplest, the water or hydrologic cycle can be broken down into three processes: evaporation, condensation, and precipitation. The Sun's energy causes liquid water to evaporate—that is, it turns from a liquid to a gas (water vapor). Because water vapor is less dense than molecular nitrogen and oxygen, the major components of the atmosphere, humid air tends to rise. As it does so, it cools, eventually condensing into a water droplet. These water droplets often coalesce to form clouds. Within a cloud, the water droplets grow as they merge with other droplets until they become large enough to fall from the sky as precipitation, be it rain, hail, sleet, or snow. Back on the surface, this water will eventually begin to evaporate and the cycle starts again.

In practice, other phenomena can come into play. For example, ice can sometimes change directly into water vapor through a process known as sublimation, and water vapor can freeze—a process called deposition. Plants release water vapor into the atmosphere through their leaves via transpiration, which accounts for about 10 percent of the water vapor in the atmosphere. Water vapor can also be directly transferred to the land through condensation—as dew.

The water cycle exists because the Earth's orbit places it within the so-called Goldilocks zone, at just the right distance from the Sun and with just the right period of rotation to keep the mean surface temperature at about 57–59°F (14–15°C). This means that water can exist on the Earth in all three of its phases: as a solid, a liquid, and a gas.

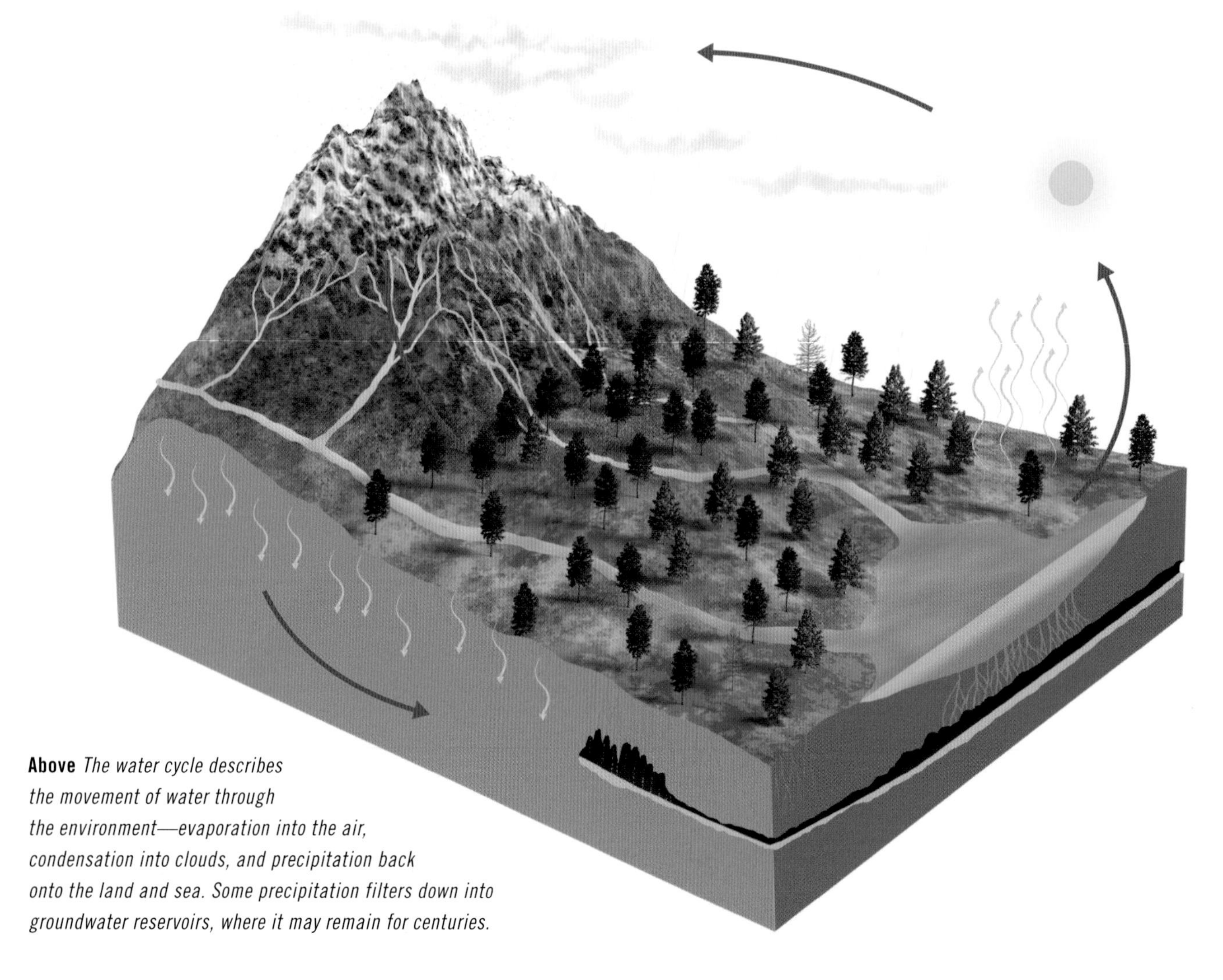

Above *The water cycle describes the movement of water through the environment—evaporation into the air, condensation into clouds, and precipitation back onto the land and sea. Some precipitation filters down into groundwater reservoirs, where it may remain for centuries.*

Each year, about 120,000 cubic miles (500,000 cubic kilometers) of water falls as precipitation, around 80 percent of it over the oceans. Estimates suggest that about 86 percent of the evaporated water that goes into the water cycle comes from the oceans. The amount of water in the atmosphere at any given time is approximately 3,100 cubic miles (12,900 cubic kilometers)—enough to cover the Earth's surface to a depth of only 1 inch (2.5 centimeters). On a global basis, evaporation roughly equals precipitation, so the amount of water vapor in the atmosphere remains approximately constant over time.

The water cycle plays a central role in the movement of heat through the world's climate system. Evaporation leads to a cooling of the nearby area, while condensation causes heating. These heat exchanges have a significant effect on the climate. Because most evaporation takes place over the surface of the ocean, it cools the surface waters. The latent heat is transferred to the water vapor and then released when it condenses.

Clouds also play a major role in regulating the Earth's climate system, either reflecting or absorbing incoming sunlight. And evaporation purifies water, as any substances present in the water are left behind when it turns into vapor.

The different places in which water spends time—the atmosphere, the ocean, the ice caps, aquifers, lakes, rivers, and so on—are known as reservoirs. The amount of time that water spends in the different reservoirs varies considerably: it only stays in the atmosphere for about nine days but can spend 10,000 years in an aquifer and some Antarctic ice has been dated to 800,000 years ago.

IMPACT OF CLIMATE CHANGE

The water cycle plays a central role in both the causes and the effects of climate change. Water vapor is the Earth's main greenhouse gas and the cycling of water is intimately linked with the energy exchanges between the atmosphere, ocean and land that determine the Earth's climate. Global warming is already changing the way in which that cycling is taking place, altering rates of evaporation, precipitation patterns, cloud formation and the melting of ice. In particular, precipitation is becoming more intense in some places and much more infrequent in others, leading to an increase in floods and droughts. Elsewhere, more precipitation is falling as rain rather than snow, speeding up the retreat of glaciers, which are already melting due to an increase in air temperature.

Below *Rain falls from a cloud over the English countryside during summer.*

// Clouds

Vast, shifting collections of tiny water droplets and ice crystals, clouds play a vital role in the water cycle and help to facilitate life on Earth by offering protection from the Sun's intense heat during the day and acting as a blanket to keep us from getting too cold at night.

Clouds form when water vapor condenses into visible water droplets or ice crystals. This will only take place if the air is saturated—that is, it is holding as much water vapor as possible. The air's temperature determines how much water vapor it can hold: in general, the warmer the air, the higher its capacity to hold water vapor. Thus saturation is reached either by increasing the air's water content through a process such as evaporation to the point where it cannot hold any more, or by cooling the air so that it reaches what is known as its dew point, below which condensation takes place. The altitude at which the dew point is reached and clouds form is called the condensation level.

Typically, clouds appear when air rises in the lower part of the atmosphere and expands due to the drop in atmospheric pressure. The energy loss from the air due to expansion causes it to cool. For every 330 feet (100 meters) the air rises, it will cool by about 2°F (1°C), depending on its humidity; moist air may cool more slowly. So, the vertical ascent of air reduces its ability to hold water vapor, causing condensation to occur. However, the water vapor still needs something to condense around. Minute particles of dust, smoke and sea salt, known as condensation nuclei, play this role.

The water droplets in a cloud have diameters of about 0.0004 inches (0.01 millimeters); within a cloud, each 35 cubic feet (1 meter) of air will contain about 100 million droplets. Because they are extremely small, droplets can remain in liquid form at temperatures as low as -22°F (-30°C). When they remain liquid despite being below their normal freezing point, they are called supercooled droplets. Their size also makes them extremely light, allowing them to remain aloft.

Most clouds form within the troposphere; however, they do sometimes form in the stratosphere and mesosphere. Polar stratospheric clouds come into being in the lowest part of the stratosphere during winter, at an altitude of about 9–16 miles (15–25 kilometers), while polar mesospheric clouds, known as noctilucent

A mixture of cumulus humilis and cumulonimbus clouds over the Great Dividing Range in East Gippsland, Victoria, Australia.

clouds, form at an altitude of about 50–53 miles (80–85 kilometers). Clouds that form high in the troposphere, where temperatures are extremely low, are composed of ice crystals that can be about 0.004 inches (0.1 millimeters) in length.

Clouds are white because the droplets within them scatter all the colors of the light spectrum by about the same amount, so white light from the Sun emerges from the clouds still white. They turn grey as they become thick enough to prevent light from passing through them, creating a shadowy appearance. As droplets coalesce and grow within a cloud, the space between them increases, permitting light to penetrate farther into the cloud. The droplets also become better at absorbing light, so less is scattered back out, making the cloud look darker.

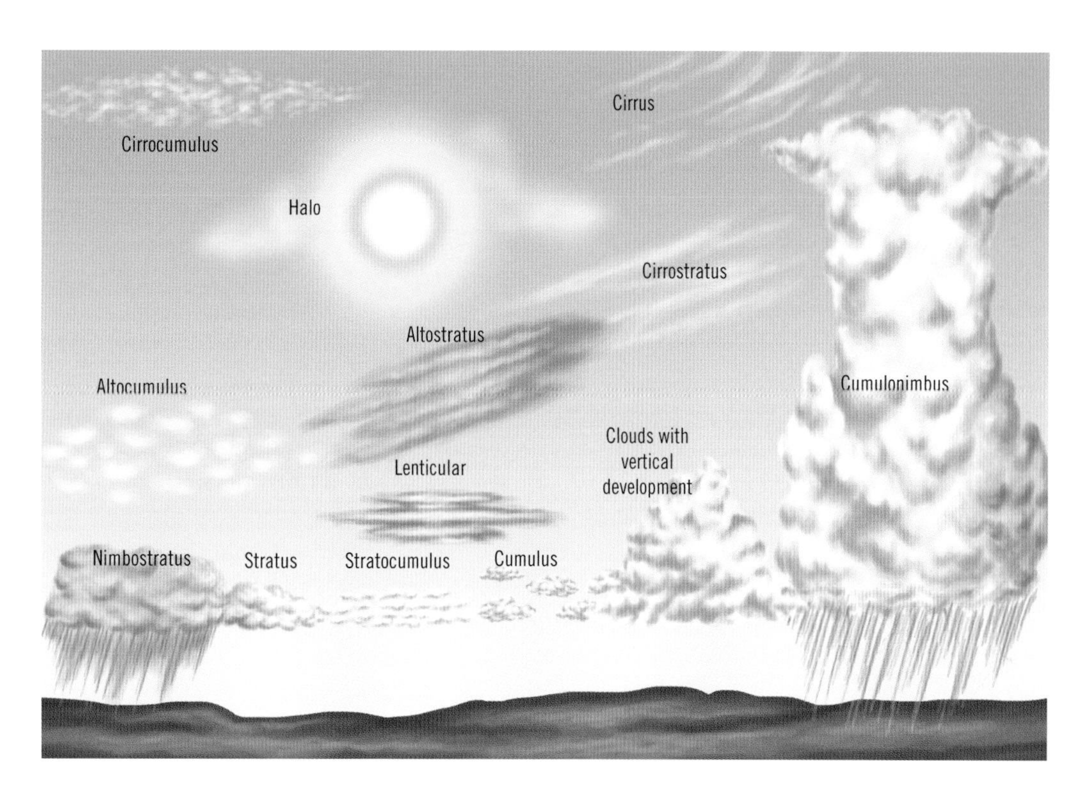

Above *Clouds are typically divided into categories based on their physical forms and altitudes. The four main cloud types are cumulus (heaped or piled up), stratus (in a layer), cirrus (thread-like, hairy or curled), and nimbus (rain bearing).*

// Precipitation

Water falling from clouds, whether liquid or frozen, is called precipitation and plays a major role in the global water cycle.

Precipitation takes numerous forms, including drizzle, rain, sleet, snow, ice pellets, ice needles, graupel, and hail. Sometimes, different types fall simultaneously.

Rain is the most common form. It is usually measured in inches (or millimeters) using a rain gauge; 0.04 inches (1 millimeter) of rainfall is equivalent to about 2 pints per 11 square feet (a liter of water per square meter). Snowfall is usually measured in inches (or centimeters).

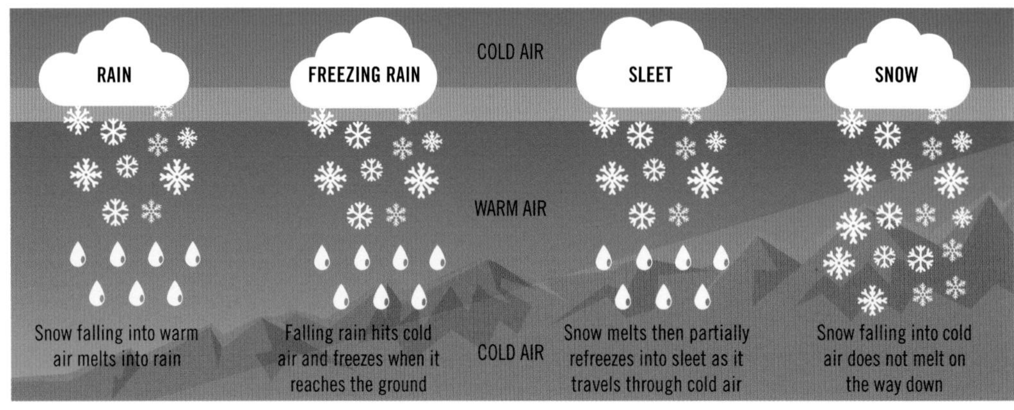

Above *The form taken by winter precipitation depends largely on the temperature of the air beneath the clouds. If the air is warm, falling ice crystals will melt to form rain, but if there is a layer of cold air below the warmer air, it may refreeze to form sleet or freezing rain. Only if the air is cold enough will it stay frozen the whole way down and fall as snow.*

About 120,000 cubic miles (505,000 cubic kilometers) of water falls as precipitation each year. This averages out to a global annual precipitation of 39 inches (990 millimeters); over land, it is 28 inches (715 millimeters).

The intensity and duration of a rainfall event are usually inversely related: short bursts of heavy rainfall or longer periods with lighter rain. Short, intense rainfall in scattered locations is known as showers. In any given place, a significant part of the annual precipitation falls on only a few days—often about half during the 12 days with the most precipitation. Areas in the path of a tropical cyclone can receive a year's worth of rainfall in just a few days.

Below *Hail forms when water droplets and ice crystals are repeatedly circulated within cumulonimbus clouds. Borne upwards by strong updraughts, they accrete layers of ice, then fall back down and the process begins again. Eventually, they become too heavy and fall to earth.*

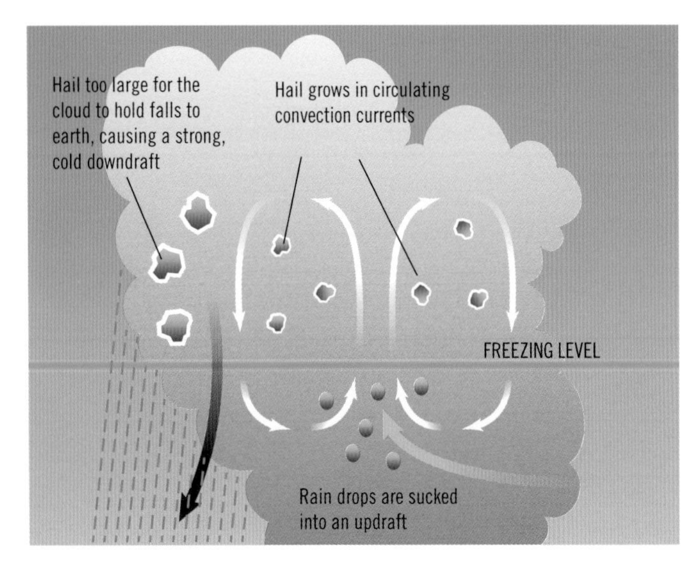

Precipitation begins to form when turbulence causes tiny water droplets in clouds to collide with each other and coalesce into larger droplets, or when droplets freeze on ice crystals. When the weight of the droplets or crystals becomes such that they overcome air resistance, they descend, colliding and coalescing, before eventually falling as precipitation. A single raindrop may contain a million or more cloud droplets.

Hail forms in cumulonimbus storm clouds when supercooled water droplets freeze around condensation nuclei such as dust or salt particles. Updraughts send the hailstones into the cloud's upper reaches before dissipating, causing the hailstones to fall, only to be lifted again. As they undergo this up-and-down motion, the hailstones accrete layers of ice. Eventually, they become too heavy for the updraught and fall from the cloud. Hailstones have a minimum diameter of 0.2 inches (5 millimeters), but can grow to 6 inches (15 centimeters) across and weigh more than 1.1 pounds (0.5 kilograms).

Cloud droplets can stay liquid at temperatures well below freezing, a phenomenon called supercooling. However, at or below about -40°F (-40°C), they freeze spontaneously.

Above *This conceptual image shows how the size and distribution of raindrops varies within a storm cloud. Blues and greens represent small droplets (0.02–0.12 inches/0.5–3mm in diameter); yellows, oranges and reds represent larger droplets (0.16–0.24 inches/4–6mm in diameter).*

At these temperatures, water vapor also forms ice crystals through sublimation around dust particles (sublimation nuclei), and grow quickly as more water vapor is sublimated. The crystals also stick to other crystals, a process known as aggregation. Eventually, they become so heavy that they fall from the cloud. If the air below the cloud is cool enough, they fall as snow; at higher temperatures, they melt and fall as rain (indeed, most rain begins as snowflakes).

When so many water droplets freeze on the surface of a snow crystal that its original shape is no longer identifiable, the small, fragile ball of ice is known as graupel or soft hail. Graupel typically falls in place of normal snowflakes, often together with ice pellets, which are small, translucent balls of ice, smaller than hailstones, formed when a layer of above-freezing air is trapped between layers of sub-freezing air. Snowflakes falling from the top layer melt in the middle layer then refreeze into ice pellets.

Raindrops range in size from 0.004 to 0.35 inches (0.1–9 millimeters) in diameter, although those more than 0.2 inches (4.5 millimeters) across tend to split into smaller drops as they fall. They do not resemble teardrops: those up to about 0.04 inches (1 millimeter) across are almost perfectly spherical; air resistance causes larger drops to flatten out, with a dimple on the underside.

There are three main types of precipitation: convective, stratiform, and orographic. Convective rainfall is created by strong vertical motion of the air and typically causes heavy showers over a limited area. Stratiform processes involve weaker upward motion and the rainfall tends to be less intense. Orographic precipitation occurs when moisture-laden winds run into steep mountains. The air is forced upward and rapid cooling causes the moisture to condense and fall as rain on the mountain's windward side. The area on the mountain's leeward side is said to exist in a rain shadow as the winds that blow over it have no moisture left.

Liquid precipitation made up of very small drops (diameters 0.008–0.02 inches/0.2–0.5 millimeters) is termed drizzle. It forms when cloud droplets coalesce in low clouds that contain only weak updrafts, and falls when the relative humidity below the cloud base is high; otherwise, the drops evaporate before they reach the ground.

The term sleet is used to mean ice pellets in the USA and a mixture of snow and rain in the UK and most Commonwealth countries.

Sometimes, most commonly over deserts, precipitation evaporates before it reaches the ground. This is called virga.

ACID RAIN

Although all precipitation is freshwater, atmospheric pollutants—in particular sulfur dioxide and nitrogen oxides—can contaminate water droplets before they fall to Earth. The pollutants react with the water to form weak acids, hence the droplets are called acid rain. Plants, aquatic ecosystems, and even bridges and buildings can suffer adverse effects from acid rain.

Below *Hailstone formation within a thunderstorm.*

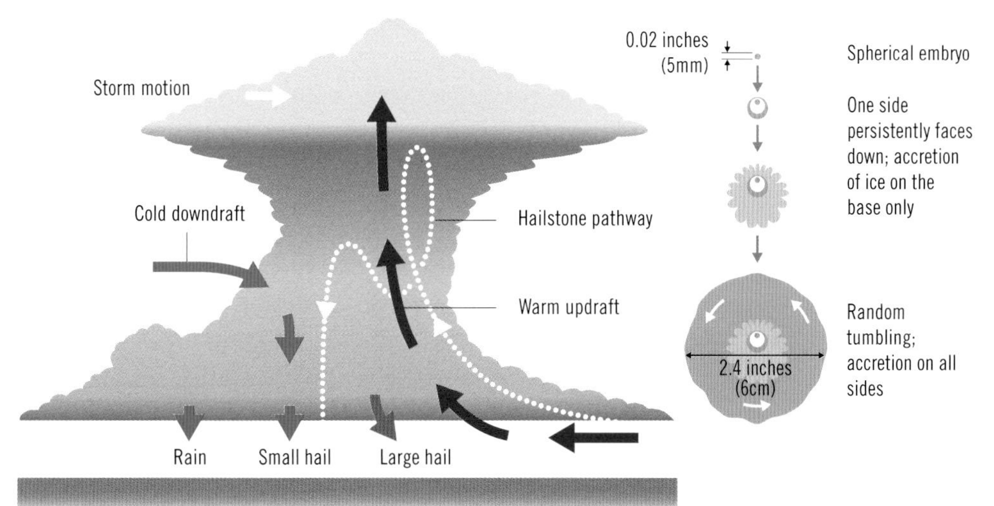

// Global precipitation patterns

Averaged over the whole Earth, the mean annual global precipitation is about 40 inches (1,050 millimeters); however, its distribution is extremely patchy.

Local factors, such as the presence of mountain ranges, can have a significant effect on the amount of precipitation that falls in a given region, but broad geographical patterns are also evident. These are the result of the spatial distribution of global air-pressure systems, winds, water availability through evaporation, and mechanisms for causing air to rise. Rising air means more precipitation; falling air means less.

On a global level, the highest rainfall is found in the tropics and the monsoon area of South and Southeast Asia (see page 172). The region around the equator receives high rainfall because of the impact of what is known as the Inter-Tropical Convergence Zone (ITCZ). Here, the warm, moist trade winds (see page 168) from both hemispheres converge, creating a general upward flow of air. In addition, heating of the land causes frequent thunderstorms, which produce considerable amounts of rain.

Moving into the subtropics, rainfall becomes much more variable. A band of high atmospheric pressure sits over these regions, causing the air to fall and become warm and dry. This is largely responsible for the ribbon of deserts that encircles the planet at around 20° of latitude. Along the east coasts of the subtropical landmasses, rainfall is much higher than on the west coasts, as the easterly trade winds, which have been picking up moisture as they blow over the warm oceans, run up against coastal mountains.

In general, the mid-latitudes receive moderate amounts of precipitation, with rainfall dominated by traveling depressions and fronts. The collision of large air masses with very different temperatures—warm, moist air from the subtropics bumping into cold, dry polar air—regularly causes heavy rainfall.

At high latitudes, and particularly at the poles, there is little evaporation from the cold surfaces of the sea and land, and the cold air cannot hold very much moisture, so precipitation is relatively rare. This is compounded by the fact that high-latitude belts of high air pressure cause the air to fall. Hence Antarctica is considered to be a cold desert. In the high latitudes, western coasts are usually wetter than eastern coasts.

Global precipitation patterns also show seasonal changes: as the amount of incoming radiation from the Sun changes, so too do the movements of air masses and hence rainfall patterns. Over the equator, the ITCZ moves towards the hemisphere that is experiencing summer, which brings rain, with a correspondingly dry season in winter. The high-pressure band that sits over the subtropics also tends to shift with the seasons, bringing dry summers on the poleward side and dry

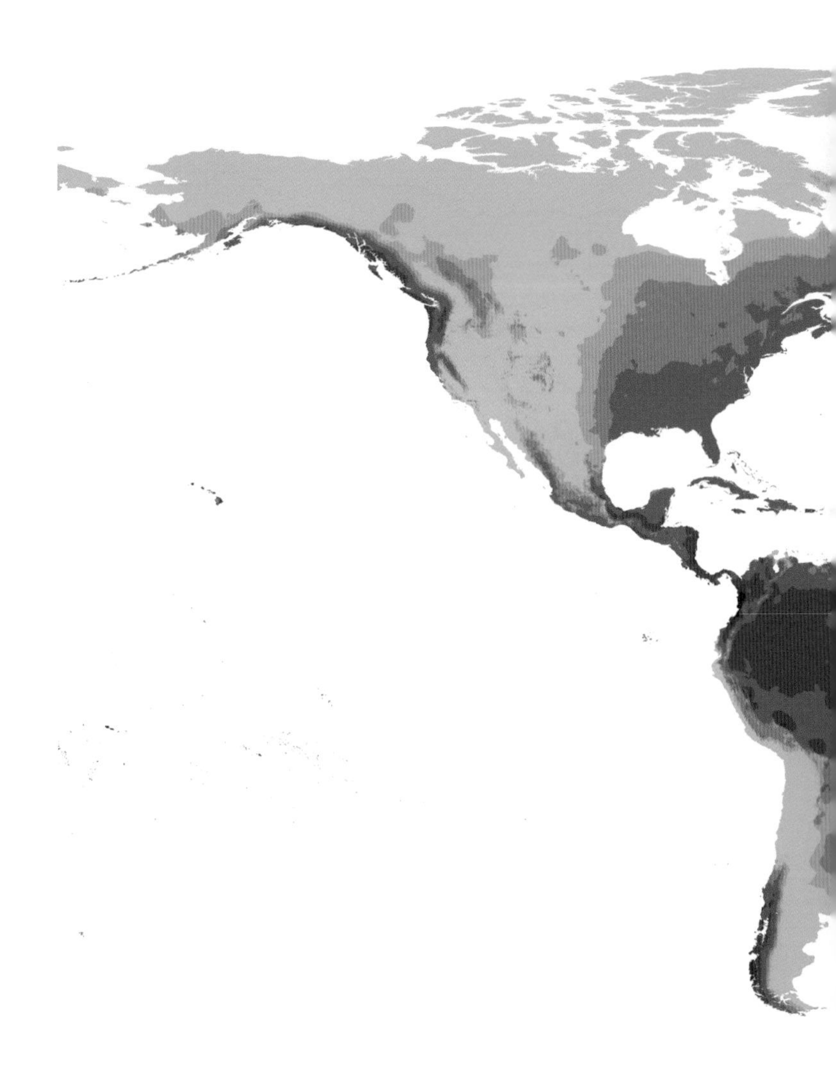

Above *Global mean precipitation. In general, precipitation tends to be higher in the tropics and along the coastal margins of the continents, while there are broad bands of low precipitation at the high and mid-latitudes, which correspond to areas of high atmospheric pressure.*

winters on the equatorward side. In general, areas mostly north of the equator receive their rain from November to April, while areas south of the equator receive theirs between May and October.

One of the most significant factors in rainfall patterns is the El Niño-La Niña cycle (or El Niño-Southern Oscillation, ENSO, see page 164), which affects both the intensity and the location of rainfall. Indeed, ENSO is the single biggest factor that explains the temporal variations in global rainfall.

The presence of large urban areas can also have an impact on rainfall. A phenomenon known as the urban heat-island effect causes cities to be several degrees warmer than the surrounding areas. This makes the overlying air rise, which can lead to additional shower and thunderstorm activity. In some cases, the amount of rain that falls downwind of cities may be double that which falls on the upwind side.

The margins of continents are usually fairly wet, as onshore winds bring moist air from the oceans. The interiors of continents are usually relatively dry because they are far away from sources of moisture.

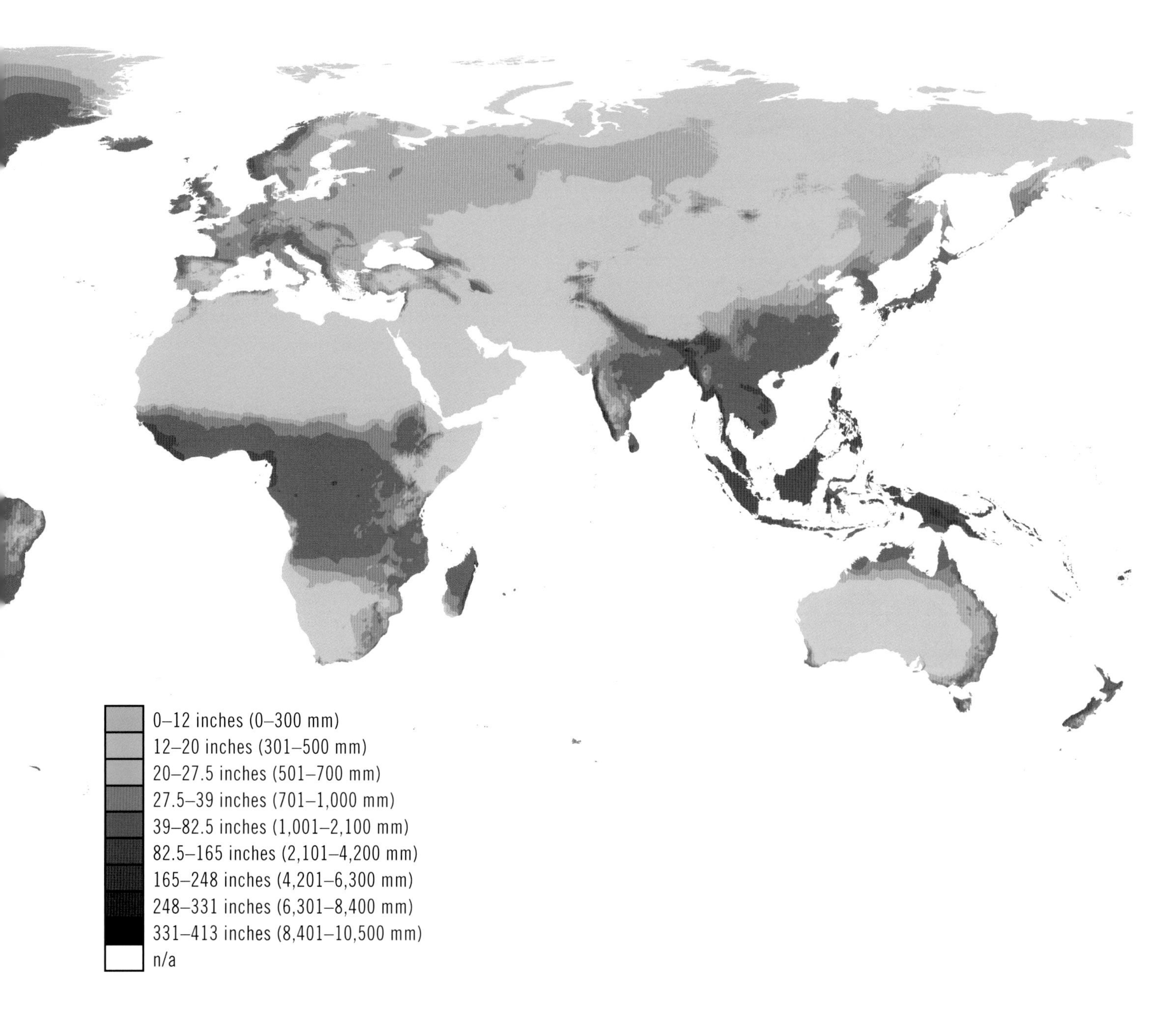

// Storms

Potentially devastating to life and property, storms take many forms.

Above *A tornado descends from a cumulonimbus cloud in Colorado, USA. On average, there are 3.5 tornadoes per 10,000 square miles (1.4 per 10,000 square kilometers) per year in the USA. Most occur in a central and southeastern region colloquially known as Tornado Alley.*

Storm is a general term used to describe violent weather conditions. Hence it covers a wide range of different phenomena, primarily involving heavy rain and strong winds. Storms can be brief, lasting mere minutes, or settle in for several days.

Most storms form when an area of low atmospheric pressure runs up against an area of high pressure, creating strong winds and storm clouds such as cumulonimbus that bring heavy precipitation. The most common are thunderstorms (see page 156), which, in addition to generating lightning and thunder, often bring strong winds, heavy rain, and/or hail. They can also spawn extremely powerful but short-lived windstorms (see below) known as microbursts. When a group of fast-moving, severe thunderstorms sweeps over an area of land it can generate a widespread, long-lived, straight-line windstorm known as a derecho.

The most powerful thunderstorms, known as supercell thunderstorms, can spawn tornadoes—destructive whirlwinds that appear as a dark, twisting, funnel-shaped cloud that reaches from the ground to the base of a cumulonimbus cloud. In the most extreme cases, tornadoes can exhibit wind speeds of more than 300 mph (480 km/h), measure 2 miles (3 kilometers) across and travel more than 60 miles (100 kilometers). If a tornado forms over water, it is known as a waterspout. Multiple waterspouts sometimes occur at the same time in a small area.

When the ocean surface warms, it can lead to the formation of cyclones, large low-pressure systems surrounded by strong, inward-spiraling winds (see page 158). The strongest cyclones form in the tropics. Occasionally, a mid-latitude cyclonic low-pressure area will suddenly deepen, becoming what is known as a bomb cyclone. Winds within such a storm can be as powerful as those in a tropical cyclone. Extratropical storms that exhibit sustained wind speeds in the range of 30–60 mph (50–100 km/h) are known as gales (although exact criteria vary between different meteorological bodies).

Storms that feature strong winds but little or no precipitation are termed windstorms. Occasionally these will pick up loose soil to form dust storms or, more rarely, sandstorms, which can be so dense that sunlight is blocked out and it becomes as dark as night. Measurements taken in a Chinese dust storm in 2001 suggested that it contained 6.5 million tons (5.9 million tonnes) of dust and covered an area of 520,000 square miles (1.34 million square kilometers). If winds suddenly increase significantly for at least a minute, it is known as a squall.

Cold conditions bring other types of storm. When below-freezing air close to the land surface is overlaid by a thick layer of warmer air, falling rain freezes when it lands, creating a thick layer of ice. These so-called ice storms can bring down tree limbs and powerlines, and make driving very dangerous. Extremely heavy snowfall over an extended period is known as a snowstorm. When heavy snowfall is accompanied by gale-force winds and extreme cold, it is called a blizzard.

At the other extreme, wildfires that become so intense that they create and sustain their own wind systems are known as firestorms.

The likelihood that a particular type of storm will take place is known as the return period or frequency. A storm with a 10 percent likelihood of occurring in any given year is called a one-in-ten-year storm. The longer the return period, the more extreme the storm—so a one-in-ten-year event will be less intense than a one-in-100-year event.

Below *A supercell storm in Oklahoma, USA. A supercell is the most powerful type of thunderstorm, characterized by the presence of a deep, persistently rotating updraft, known as a mesocyclone.*

Top: *A fire whirl or fire devil rises out of intense flames during the Pine Gulch Fire near Grand Junction, Colorado, in August 2020. Short-lived whirlwinds that are spawned by the intense heat and turbulent winds created by a wildfire, fire whirls can be up to 0.6 miles (1 kilometer) tall and reach temperatures of more than 1,832°F (1,000°C).*
Above *A dust storm approaches Phoenix, Arizona, USA.*

// Thunderstorms

Producing heavy rain or hail, strong, gusty winds and spectacular light shows accompanied by powerful explosions of sound, thunderstorms are usually the result of collisions between warm and cold air masses.

Thunderstorms typically form when cool, heavy air moves over warmer, lighter air. As the cooler air sinks, it causes the warmer air to rise. If the warm air contains a large amount of moisture, it will condense and form huge, towering cumulonimbus clouds whose bases are usually around 1 mile (1 to 2 kilometers) above the ground and whose tops can be up to 9 miles (15 kilometers) high. As the water vapor condenses, it releases latent heat energy, which fuels the upward motion of the air.

At the center of the thundercloud, air is rising rapidly and temperatures range from +5 to -13°F (-15°C–-25°C). These conditions produce a mixture of super-cooled (below-freezing) water droplets, small ice crystals, and graupel (soft hail). The water droplets and ice crystals are carried upwards on the updraft; being larger and denser, the graupel tends to remain where it is or fall slightly, resulting in constant collisions with the rising ice crystals. These collisions impart a positive charge to the crystals and a negative charge to the graupel. The positively charged ice crystals continue to rise until they reach the top of the storm cloud, while the graupel either remains at the same height or drops towards the cloud base. This results in the cloud's upper regions becoming positively charged, while

Below *The main components of a thundercloud. Thunderstorms are driven by strong updrafts of warm, moist air, which circulate water droplets and balls of ice, causing them to become electrically charged.*

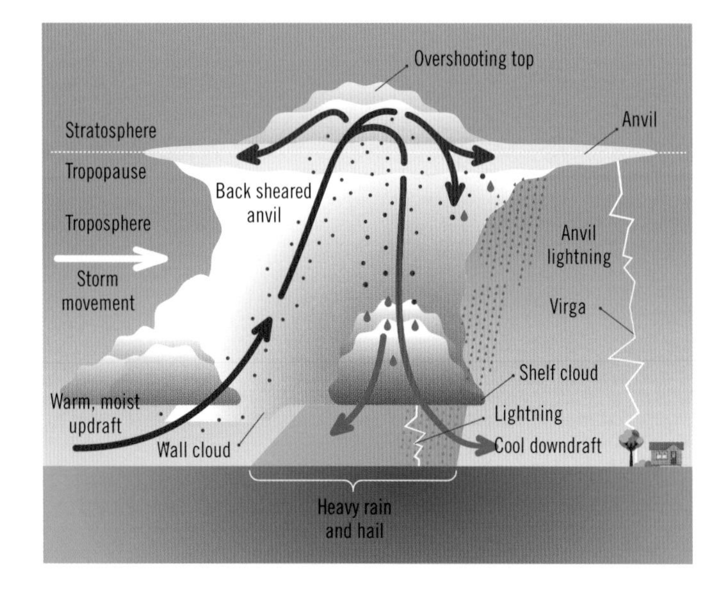

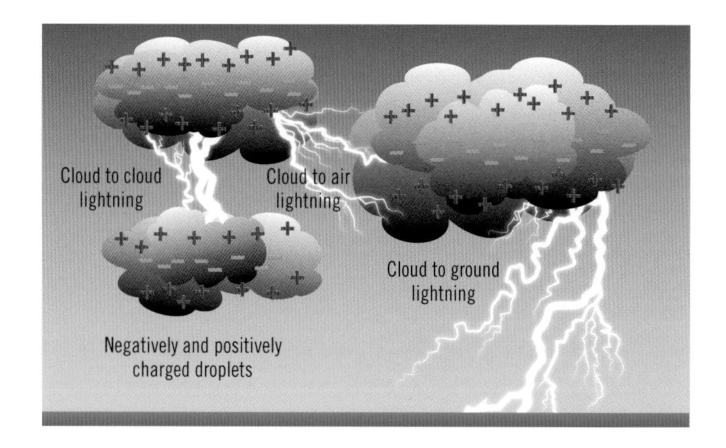

Above *The distribution of charged droplets within thunderclouds is what causes lightning discharges to take place.*

the middle and lower sections take on a negative charge. When the accumulated electric charges become large enough, a lightning discharge takes place.

Lightning can occur within a thundercloud, between two different clouds, or between a cloud and the ground. Only about a quarter of lightning flashes are from a cloud to the ground; flashes within a cloud (the most frequently occurring type) or between two clouds are more common.

Most lightning strikes result in the instantaneous release of about one gigajoule of energy. They typically last for 0.2 seconds, but consist of a number of much shorter flashes (strokes) of around 60–70 microseconds.

Around the world, cloud-generated lightning occurs about 45 times per second, which equates to almost 1.4 billion flashes per year. About 70 percent occurs over land in the tropics. Records suggest that lightning strikes most often in eastern Democratic Republic of the Congo near the small mountain village of Kifuka, where 410 lightning strikes occur per square mile (158 per square kilometer) per year. Lightning can also be generated during dust storms, forest fires, tornadoes, and volcanic eruptions.

Thunder is caused by the rapid heating of air by a bolt of lightning. During a lightning flash, the narrow channel of air through which the lightning passes is suddenly heated to temperatures of up to 54,000°F (30,000°C), causing the air to expand rapidly. This creates the rippling shockwave that we hear as thunder.

When a lightning flash is close, the thunder will be heard

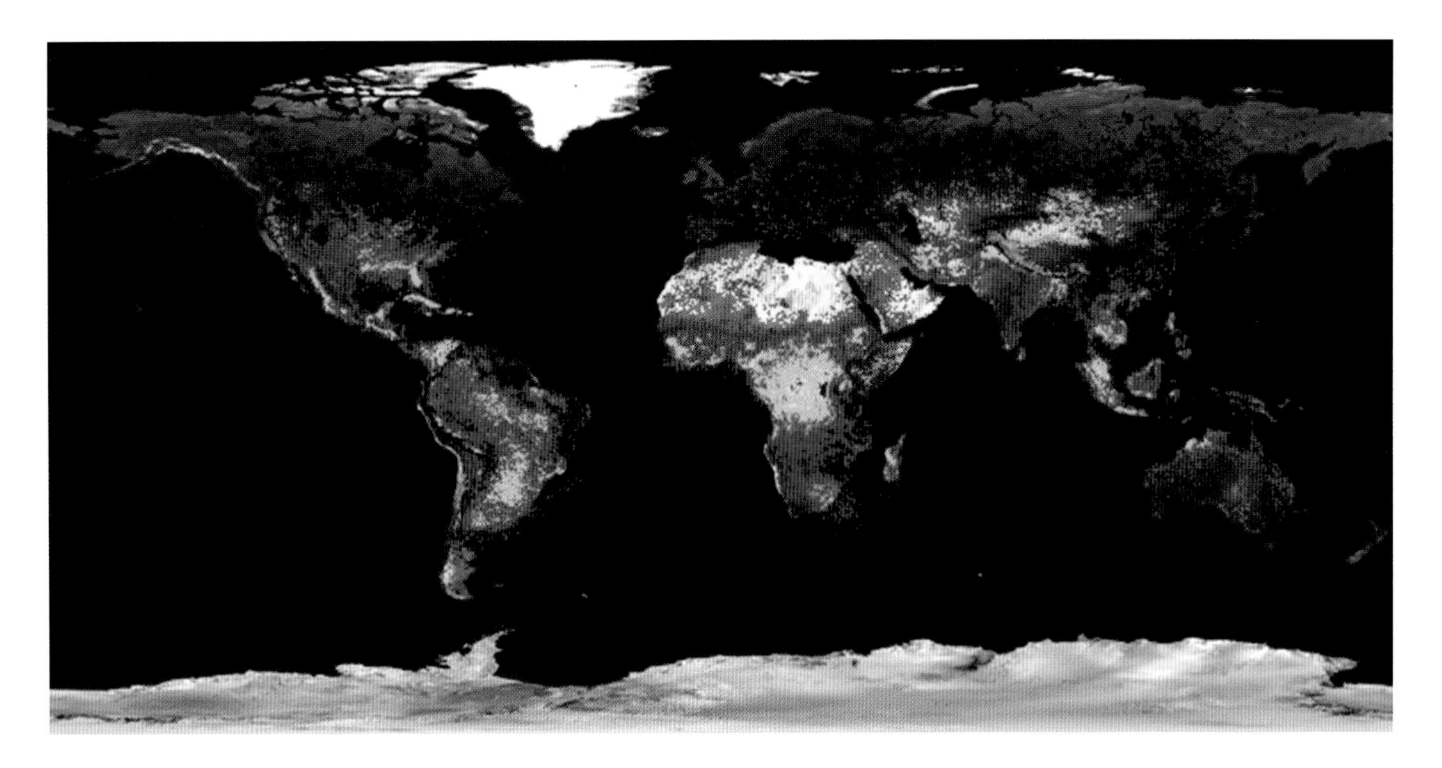

Above *This map shows the number of lightning flashes per square kilometer per year from 1995 to 2002, ranging from deep blue (fewer than ten) through to deep red (more than 100). Clearly visible is the region in Central Africa that receives an extraordinarily high rate of lightning flashes. But also note the Brahmaputra Valley in far eastern India, which experiences the world's highest monthly lightning flash rate between April and May as the annual monsoon is starting, and the region around Venezuela's Lake Maracaibo, where 650 lightning flashes occur per square mile (250 per square kilometer) each year.*

as a sudden loud crack; more distant thunder is heard as a low, long, drawn-out rumble because the sound waves are created along the length of the lightning, so they reach the observer at different times. Because light travels significantly faster than sound, thunder is heard after the lightning that generated it is seen.

You can work out roughly how far away the lightning is in kilometers by counting the seconds between the flash and the associated thunder, and dividing by three (for miles, divide by five). Lightning can be seen from more than 100 miles (160 kilometers) away, whereas thunder only travels about 15 miles (25 kilometers), so sometimes lightning can be seen but no thunder heard.

Lightning sometimes occurs in the absence of rain. This "dry lightning" is the most common natural cause of wildfires.

Below *A series of cloud-to-ground lightning strikes during a thunderstorm in Aguila, Arizona, USA.*

// Cyclones

Among the strongest and most intense storms on Earth, cyclones cause millions of dollars of damage around the world every year.

Left *Hurricane Florence moves across the Atlantic Ocean, as seen from the International Space Station on 12 September 2018. Originating near Cape Verde, the storm brought heavy rainfall and flooding to the eastern USA and spawned ten tornadoes in the US state of Virginia.*

A cyclone is a large system of inward-spiraling winds that rotates around an area of low atmospheric pressure (the term cyclone comes from the Greek word *kuklos*, which signifies, among other things, the coil of a snake). Cyclones rotate anti-clockwise in the Northern Hemisphere and clockwise in the Southern Hemisphere. (Comparable weather systems that form around areas of high pressure are known as anticyclones. These systems rotate in the opposite direction, feature weaker winds and tend not to produce precipitation.)

Cyclones mostly occur in the mid- and high latitudes. Because they need the Coriolis effect (see page 106) in order to circulate and the effect is absent at the equator, they almost never occur in equatorial regions and never cross the equator itself.

There are two main types of cyclone: mid-latitude or extratropical and tropical. Tropical cyclones are much smaller than mid-latitude cyclones: the former typically have diameters of about 125–310 miles (200–500 kilometers) although they can reach 620 miles (1,000 kilometers), whereas for the latter, the range is from nearly 620 to 2,500 miles (1,000 to 4,000 kilometers). Tropical cyclones also tend to be considerably more violent, with wind velocities that may exceed 185 mph (300 km/h), compared to a maximum of about 75 mph (120 km/h) for mid-latitude cyclones.

As their name suggests, tropical cyclones form in the tropical latitudes (specifically between 10° and 25°). During summer and autumn, large masses of humid air accumulate over warm ocean water. This air rises, creating a low-pressure cell known as a tropical depression around which thunderstorms develop. If the sea-surface temperature reaches or exceeds 82°F (28°C), the air begins to rotate around the area of low pressure. The swirling air rises and cools, and the water vapor condenses, creating clouds and releasing energy from latent heat. If conditions remain favorable, the storm will build into a tropical cyclone within two or three days.

At the center of a cyclone is an area of relative calm (known as the eye in a mature tropical cyclone), where air is rising and the regional atmospheric pressure is at its lowest. In a tropical cyclone, the eye is surrounded by a ring of intense thunderstorms known as hot towers that are arranged in what are called spiral rain bands. These storms are nourished by moisture from the warm ocean.

Tropical cyclones move according to the prevailing winds, potentially traveling as much as 500 miles (800 kilometers) in a day. Rain in a tropical cyclone can fall at a rate as high as 1 inch (2.5 centimeters) an hour, which amounts to roughly 22 billion tons (20 billion tonnes) a day. If a tropical cyclone passes over an area of cooler water or land, it weakens, often turning into a tropical depression and producing intense rain and sometimes tornadoes.

Around the world, different terms are used for tropical cyclones: in the Atlantic and Caribbean regions, they are called hurricanes; in the western Pacific and China

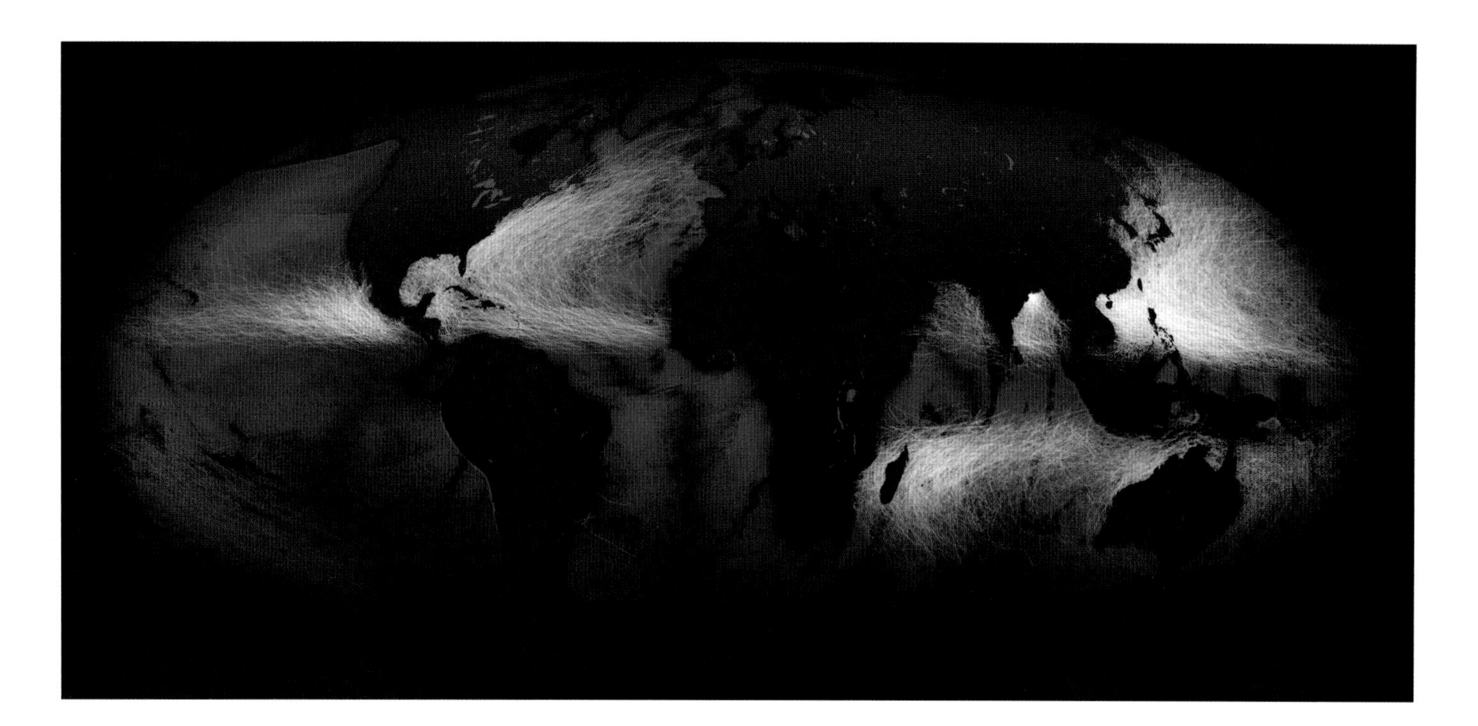

Above *Global storms between 1842 and 2017. Storm tracks are color-coded according to the ocean basins in which they formed. Brighter areas indicate regions in which a large number of storm tracks have overlapped.*

Sea, they are called typhoons. Tropical cyclones above a certain strength are given names in order to make them easy to identify in warning messages and to help increase community preparedness.

Tropical cyclones are typically categorized according to their sustained wind speed. Different scales are used in different regions. In the Western Hemisphere, the Saffir-Simpson Hurricane Wind Scale ranks tropical cyclones from category 1 (wind speeds of 74–95 mph/119–153 km/h) to category 5 (wind speeds over 156 mph/251 km/h).

Each year, about 80–100 tropical cyclones form around the world. Over the past 50 years, tropical cyclones have caused 1,942 disasters, killing almost 780,000 people and causing US$1.4 billion in economic losses. Damage can come from the high winds, heavy rainfall, coastal flooding, and storm surges, which may occur when a cyclone makes landfall, as strong onshore winds push a huge volume of water against the coast and low pressure sucks it upwards.

Below *Tropical cyclone formation. (1) Clouds form rapidly over warm water leading to updrafts that pull air in. (2) The Coriolis effect causes the system to rotate. (3) The mature cyclone features a calm central eye surrounded by a vortex of warm air and hot tower thunderstorms. Cooler air (blue arrows) can be seen flowing around and within the cyclone.*

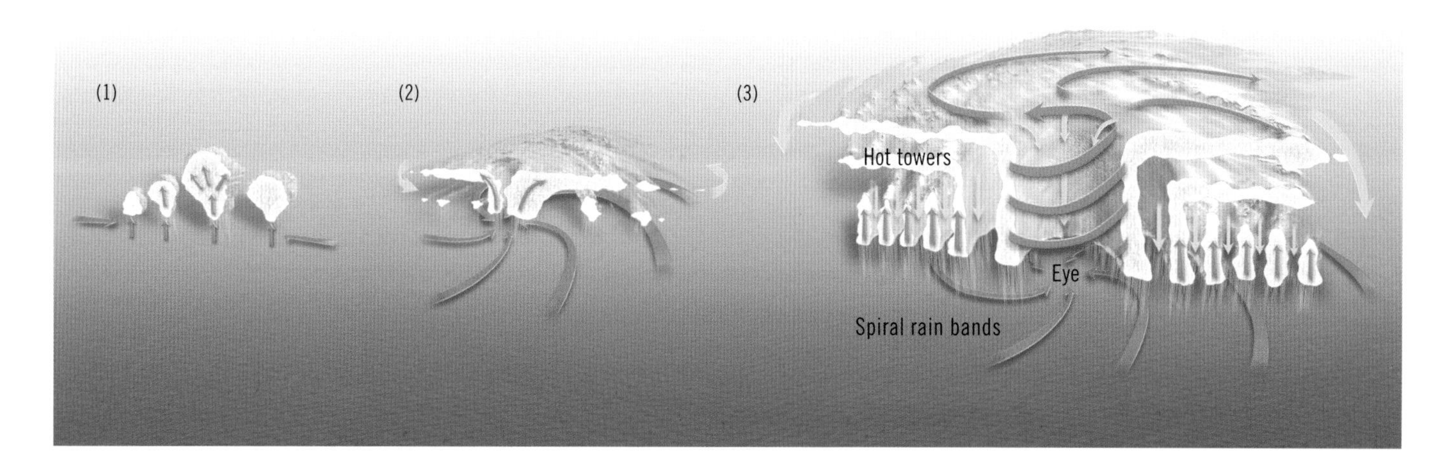

// Drought

When a region suffers a shortage of water (either precipitation or groundwater) over a prolonged period, it is said to be experiencing a drought. Recently, drought has affected more people around the world than any other type of natural disaster.

Above *An alpaca on a farm in New South Wales, Australia, during a drought. Since the 1860s, Australia has experienced a "severe" drought once every 18 years on average.*

In many parts of the globe, especially where rainfall is already low, periods of markedly reduced precipitation and consequent drought are a regular feature of the climate. In the tropics, for example, annual dry seasons, when humidity is low and surface water dries up, significantly increase the chances of a drought developing.

Droughts are generally triggered by changes to the prevailing weather patterns. Changes in air circulation, such as an alteration in wind direction that brings continental rather than oceanic air masses, potentially divert the moist air that would normally produce precipitation or simply change the way in which it is delivered. A change in the timing of rainfall or snowmelt can also throw water supply and demand out of sync, leading to shortages.

An above-average prevalence of high-pressure systems can also trigger drought by restricting the development of thunderstorms and other forms of precipitation. The El Niño-Southern Oscillation (see page 164) is often associated with drought as it shifts precipitation patterns, bringing drier conditions to a broad swathe of territory.

Drought is sometimes described as a "creeping disaster" as it can be difficult to pinpoint the beginning and end. A drought could last for anything from weeks to decades and can end as gradually as it began. The threshold for declaring drought differs from region to region, based largely on local weather patterns.

Once a region is in the grip of drought, a number of feedback mechanisms can worsen conditions. The lack of moisture means that there is no water vapor available for cloud formation, keeping temperatures high and further reducing the likelihood of precipitation. Baked soil may be resistant to rain, so any that does fall quickly flows away into rivers and streams.

Although droughts mostly occur naturally, human activity can increase their likelihood and exacerbate their effects. Over-extraction of water for irrigation can deplete aquifers; poor farming practices and the removal of vegetation can lead to dust storms or the formation of dust bowls; damming and over-extraction of water from rivers can contribute to drought in downstream regions. Land clearing can reduce the amount of water available to feed the water cycle, making entire regions more vulnerable to drought, and diminish soil quality, reducing the land's ability to absorb and retain water.

A lack of adequate precipitation can lead to diminished stream flow, crop damage, parched soils, and dry aquifers, all of which can contribute to a general water shortage. Drought can also reduce water quality by reducing the dilution of pollutants. The unusually dry conditions that characterize droughts can have significant negative consequences for agriculture and livestock farming, and hence food supplies. As a result, droughts in developing countries are often associated with famines. A drought can also increase the risk of bushfires.

Prolonged and recurring drought can permanently change a region's habitats. Recent droughts in the Amazon Basin have led to damaging wildfires and raised fears that the region's iconic rainforests are approaching a tipping point and may soon begin to turn into savanna or even desert, which would have catastrophic consequences for the world's climate, not to mention the biodiversity the forests hold.

Throughout history, drought has had a significant impact on human societies. There is evidence that drought provided the trigger for the exodus of early humans out of Africa around 135,000 years ago. During the modern era, droughts have caused civil unrest and regional conflicts, and mass migration that resulted in the displacement of entire populations. Famines triggered by droughts in China and the Soviet Union during the 1920s are estimated to have killed more than 8 million people.

Below *Overall water risk around the globe, including the physical quantity of water available and its quality. Twelve of the world's 17 most water-stressed countries are in the Middle East and North Africa.*

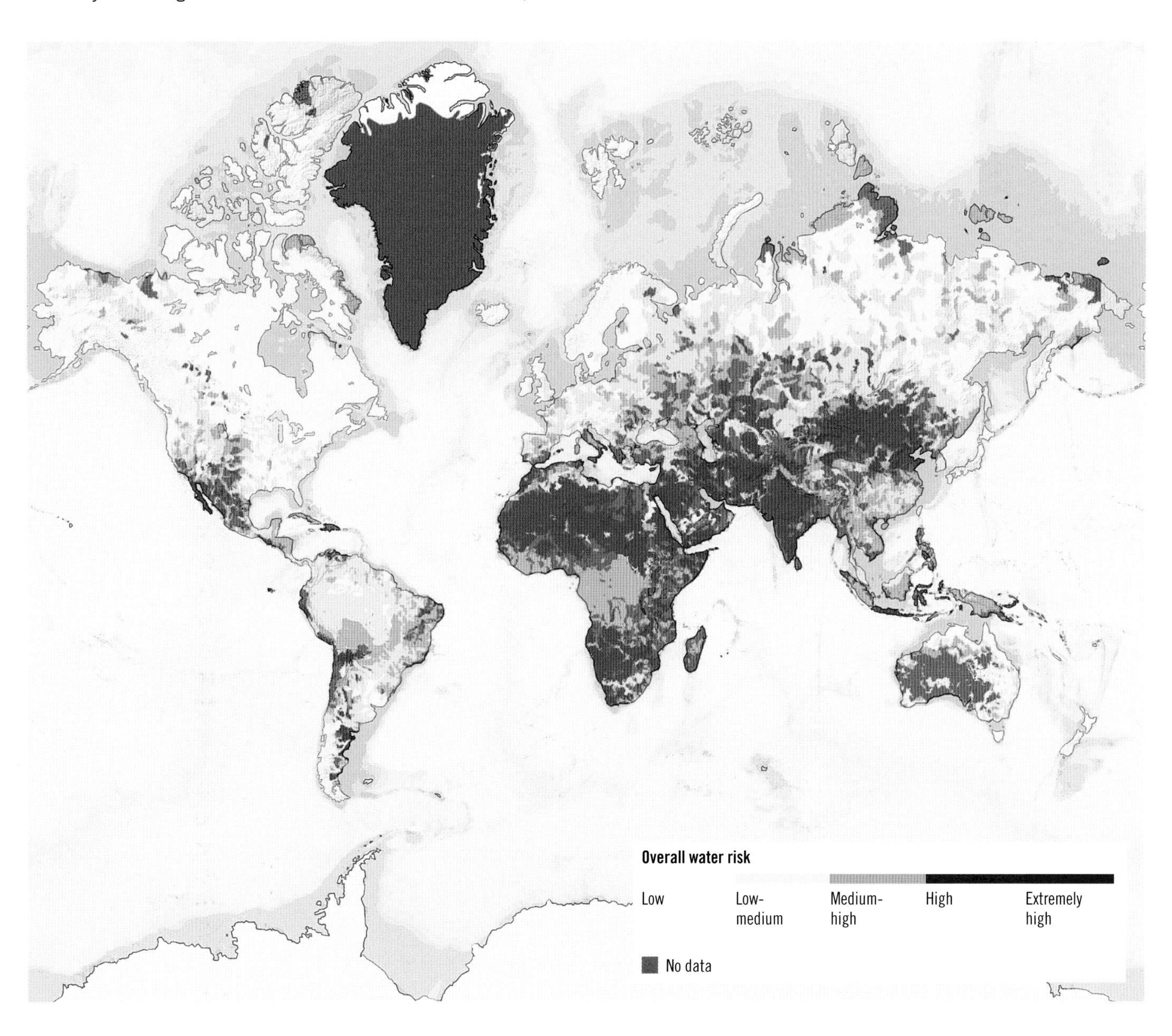

// Floods

The most common and widespread weather-related natural disasters, floods can cause devastating damage and terrible loss of life.

Floods occur when water flows out over land that is normally dry. They have numerous causes, including heavy rains, dam and levee breaches, rapid snowmelt, and inundation by the sea. They can happen rapidly, a phenomenon known as flash flooding, or build slowly and remain for weeks.

Flooding most commonly happens when water overflows from a water body such as a river or lake. If the flow rate of a river exceeds the capacity of its channel, it may break its banks and flood the surrounding area. This is particularly likely to occur at bends or meanders in the waterway. The increase in flow rate may be caused by heavy or sustained rainfall or rapid snowmelt. Extreme flood events are often the result of the coincidence of factors, such as when warm, heavy rain falls on and melts a thick snowpack. A combination of rain and snowmelt caused the deadliest flood in recorded history, in 1931 in the Yellow River valley in China, which killed an estimated 2.5–3.7 million people.

A further cause of floods may be blockage of a river channel by debris, perhaps from a landslide, causing water to back up and flow over the river's banks. Floods in urban areas may occur when rainfall overwhelms the capacity of drainage systems. Such floods may be exacerbated by paved surfaces such as streets, which prevent rainfall from infiltrating into the ground, causing higher surface run-off.

So-called areal floods may afflict flat or low-lying areas when rainfall or snowmelt arrives more quickly than it can soak into the ground or flow away. In areas where the water table is shallow, the ground can quickly become saturated and any additional water accumulates on the surface. Impermeable surfaces such as frozen ground, rock, and concrete can exacerbate areal flooding.

Most flooding-related deaths occur as a result of flash floods. Flash floods can occur due to the catastrophic failure of a natural or man-made structure that was holding back a significant amount of water. They can also take place following sudden, intense rainfall, particularly when it happens in steep, rocky, mountainous areas or in deserts, both areas where water tends to infiltrate poorly, producing rapid run-off.

Some of the most devastating flash floods occur when the barrier holding back a large reservoir fails suddenly. Glacier outburst floods, or jökulhlaups, occur when water stored in or behind a glacier is suddenly released. In 1922, the Grímsvötn outburst in Iceland unleashed about 1.7 cubic miles (7.1 cubic kilometers) of water.

Like droughts, floods are natural disasters that are often made more likely or more damaging by human activity. Deforestation, the removal of wetlands, the alteration of

Below *Flooding takes several different forms, including pluvial, fluvial, groundwater and coastal floods. Fluvial or river floods occur when rivers break their banks. Pluvial floods are independent of an overflowing water body. They occur when extreme rainfall overwhelms drainage systems in urban areas or as flash floods. Coastal flooding may be caused by storm surges—when high tides combine with low-pressure systems and heavy seas—tsunamis or tropical cyclones making landfall. Groundwater flooding occurs when the water table rises above ground level.*

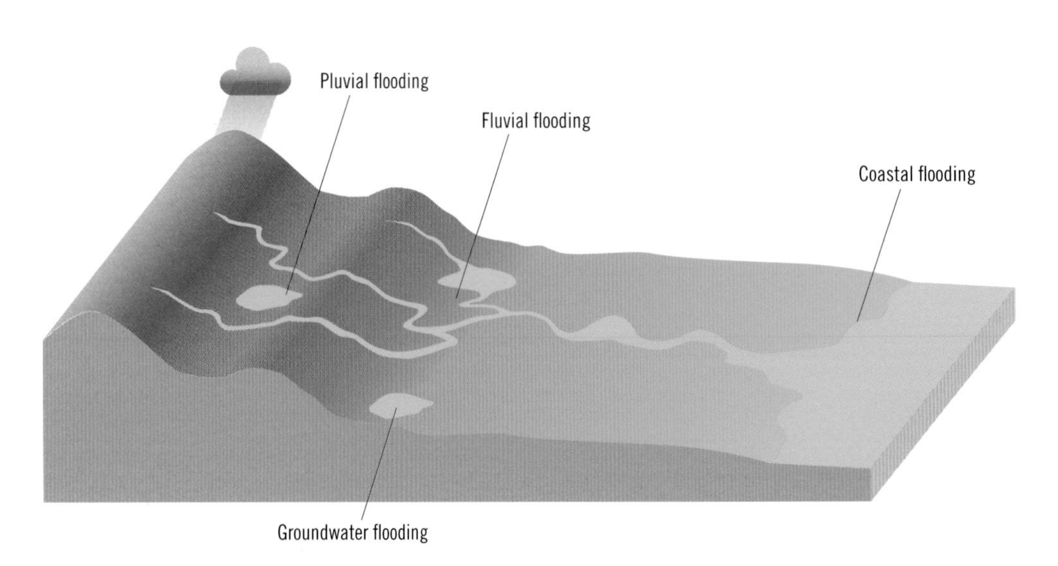

Above *Flooding of the Yangtze River in Chongqing, China, in August 2020. The 2020 summer monsoon brought regular deluges to China, with unusually strong, stationary weather systems producing frequent storms and heavy rainfall. Overall, the country experienced at least 21 large-scale floods, which destroyed crops and caused hundreds of deaths.*

river courses, and paving-over of land can all increase the frequency and intensity of floods. In recent years, flooding has become more common as urban areas have spread out across floodplains.

Floods can cause significant damage to buildings and other infrastructure, through the sheer force of flowing water or by eroding foundations. In 2007, floods in Bangladesh destroyed more than a million houses. They can also inundate farmland, destroying crops and killing livestock.

When floodwaters recede, they frequently leave affected areas blanketed in silt and mud. Waterborne diseases such as typhoid, giardia, cryptosporidium, and cholera are a particular threat in the aftermath of a flood. Water supplies are often compromised during the flood, reducing access to clean water, and human sewage may be washed into floodwaters. Stagnant floodwaters can also provide breeding grounds for mosquitoes, leading to outbreaks of diseases such as dengue and malaria.

However, floods can also have positive effects. They can be vital for communities living in river deltas as they bring nutrient-rich sediment needed for agriculture. The annual flooding of the Nile delta was a key factor in the development of ancient Egypt.

Structures built to reduce flooding include levees (embankments built along rivers to raise the level of their banks) and run-off canals, which divert water away from sensitive areas. Dams and reservoirs can also reduce the risk of downstream flooding. Wetlands on floodplains help to reduce the impact of floods by soaking up and slowing the speed of floodwaters.

// The El Niño-Southern Oscillation

Every few years, a climatic phenomenon known as El Niño kicks off in the Pacific Ocean, the impact of which is felt around the globe.

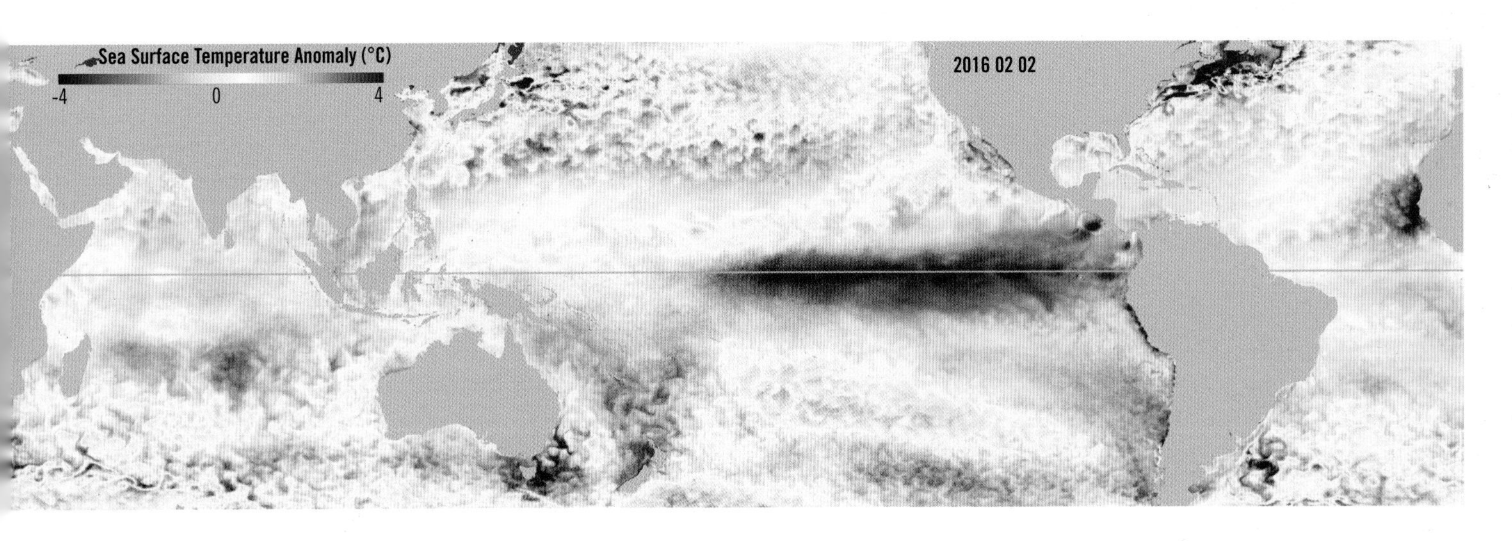

Under normal conditions, strong easterly trade winds blow away from equatorial South America, pushing warm surface water into the western Pacific. This results in an upwelling off the coasts of Ecuador, Peru, and Chile as cold, nutrient-rich water is drawn up from the deep sea, bringing bountiful conditions for fishers. Meanwhile, off Indonesia, the sea surface will be about 1.6 feet (0.5 meters) higher and 14°F (8°C) warmer than it is off Ecuador.

When an El Niño kicks in, the trade winds in the central and western Pacific weaken. This weakens the upwelling and, consequently, the surface water in the eastern tropical

Above *The sea surface temperature anomaly (the difference between the ocean temperature and the long-term average) during the 2015–16 El Niño event, one of the strongest on record. The pool of unusually warm water in the eastern equatorial Pacific is clearly visible in the center of the image.*

Below *During a normal year, the easterly trade winds push warm water towards Australasia, bringing rainfall to the region. During an El Niño event, the trade winds are weaker and the warmer water pools off South America, causing drought in Australasia. The opposite phenomenon occurs during a La Niña, with stronger trade winds pushing more warm water to the west and bringing even wetter conditions to Australasia.*

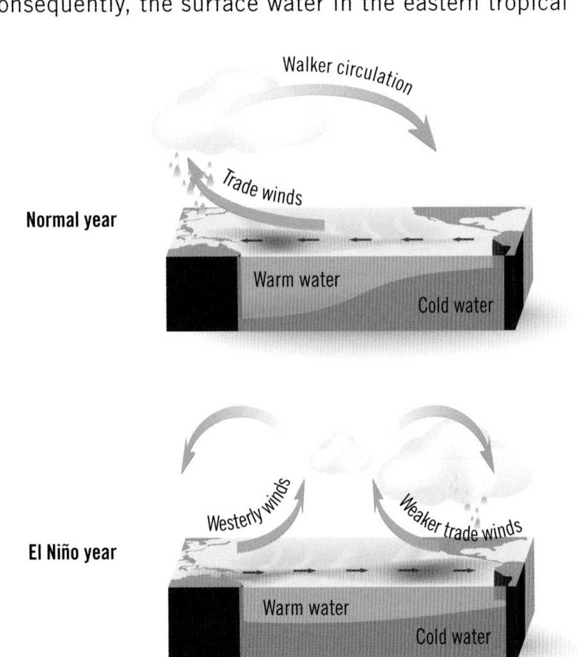

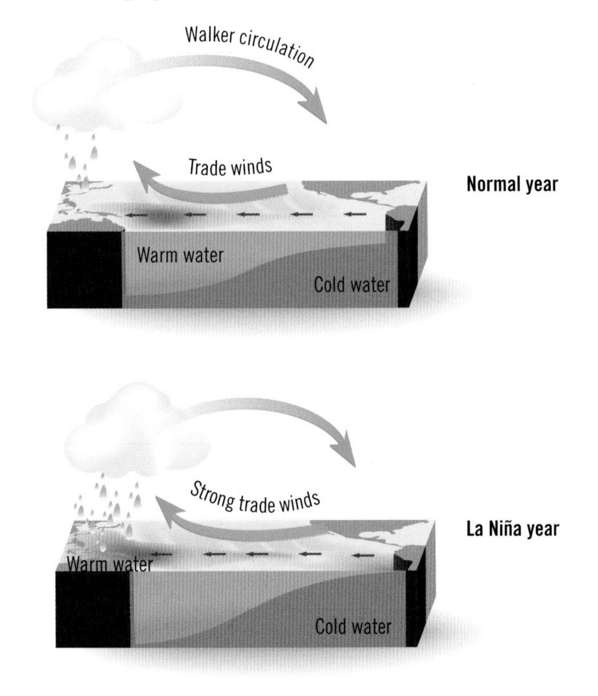

Pacific off South America warms up. This pool of warmer water feeds clouds and storms due to an increase in evaporation, causing them to shift eastwards.

During an El Niño event, the west coast of South America receives above-average rainfall and may suffer floods. Fish catches in the region are usually lower than normal, the local marine life moving north and south in search of cooler water. The shift in rainfall patterns also affects the Canadian prairies, Australia, the Pacific Islands, India and Southeast Asia, which may all experience drought conditions. Storms become less common over the Atlantic, but in the eastern Pacific, hurricanes become more common. El Niños can even increase the risk of colder winters in the UK.

An El Niño is declared when sea-surface temperatures in the tropical eastern Pacific increase by 0.9°F (0.5°C) above the long-term average for at least five successive three-month seasons; in extreme events, such as the El Niño of 1997–98, the rise can be more than 5.4°F (3°C). It is unclear what triggers an El Niño cycle, which makes the phenomenon difficult to predict.

El Niño is the "warm phase" of a larger phenomenon known as the El Niño-Southern Oscillation (ENSO). The "cool phase," known as La Niña, sees below-average sea-surface temperatures across the east-central Equatorial Pacific and an increase in the strength of the trade winds. The impacts of La Niña tend to be the opposite of those experienced during an El Niño: the tropical eastern Pacific has cooler, drier than average weather, Australia receives above-average rainfall and parts of South America may experience drought.

Although their intensity varies from episode to episode and their appearance is irregular, El Niños typically take place every three to five years or so; La Niñas tend to be less common. Both usually last for nine to 12 months, but may persist for years. They often begin to form during the Southern Hemisphere spring, reach their peak between December and January, and then weaken and disappear by May of the following year. About half of all years are neutral.

Evidence from a range of sources, including ice cores, coral and tree rings, has demonstrated that El Niño has been around for millions of years.

WHAT'S IN A NAME?

El Niños typically begin in December. Hence, during the 1600s, Peruvian fishermen named the phenomenon El Niño de Navidad—the little boy of Christmas, that is, the newborn Christ. La Niña means little girl.

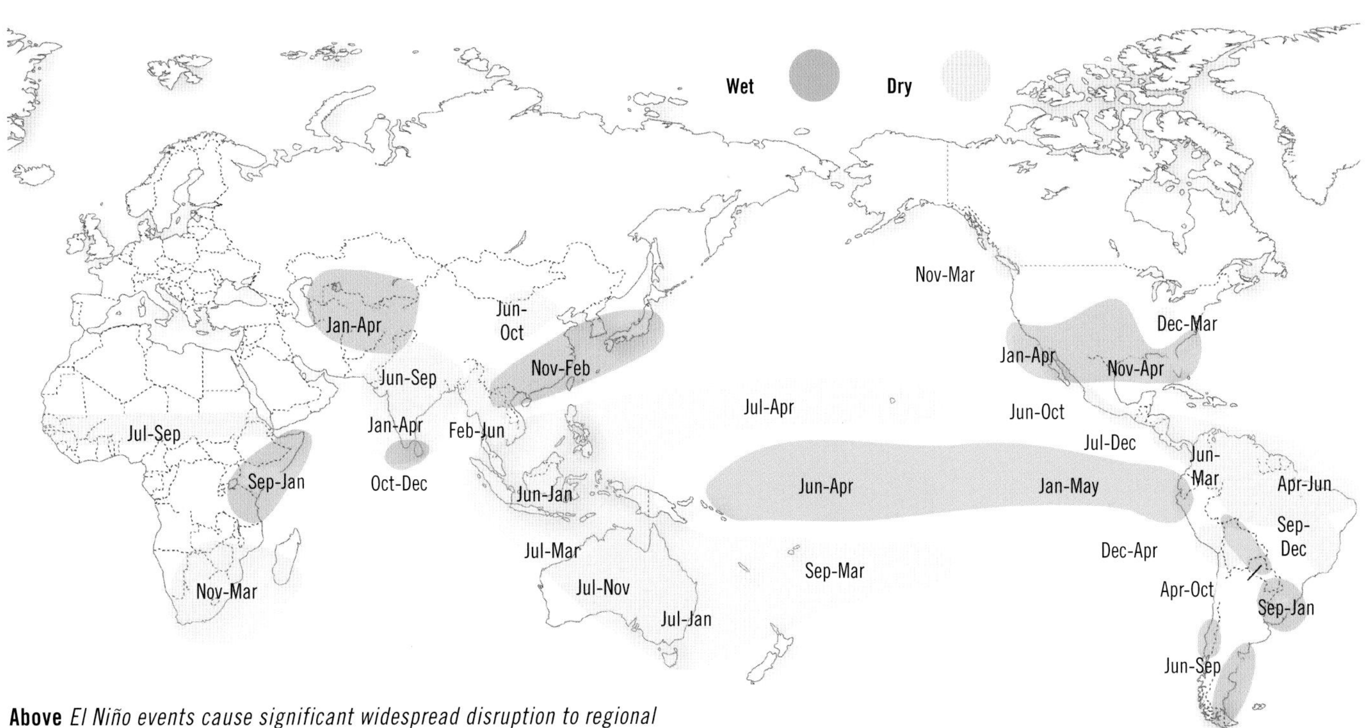

Above *El Niño events cause significant widespread disruption to regional rainfall patterns, bringing wetter conditions to some areas and drier conditions to others.*

// The seasons

Each year, much of the world goes through a regular pattern of climatic change, caused by the Earth's tilt.

We divide the year up into seasons—spring, summer, fall (or autumn), and winter—distinguished by characteristic climatic conditions, in particular day length, temperature, and weather patterns.

The Earth experiences seasons because the axis around which it spins is tilted at an angle of about 23.45° relative to its orbital plane. This means that as the Earth orbits the Sun, different regions receive different amounts of direct sunlight. During each hemisphere's summer, that hemisphere's tilt is towards the Sun, which is consequently higher in the sky. During winter, when the Sun is lower in the sky, sunlight must travel through more of the atmosphere, which dissipates the sunlight. The days are longer during summer, which also increases the amount of solar heating. The amount of heat coming from the Sun affects other aspects of the climate too, so the seasons are often marked by different weather patterns.

The Earth's two hemispheres experience opposite seasons: the Northern Hemisphere winter takes place when the Southern Hemisphere is experiencing summer. The two solstices—the longest (summer) and shortest (winter) days of the year, when the Earth's axis is pointed towards or away from the Sun—take place on June 20/21 and December 21/22. The spring (vernal) and autumnal equinoxes—the days with equal periods of daylight and darkness, when the Sun is directly overhead at the equator—take place halfway between the solstices, on March 20/21 and September 22/23.

Only the mid-latitudes experience the classic four-season year. At the higher latitudes, variation in the number of daylight hours becomes increasingly extreme. The polar regions experience only two seasons, summer and winter, with long periods during summer when the Sun does not set and periods during winter when it does not rise. Equatorial regions tend to experience little seasonal variation at all, with temperature and day length remaining similar year-round. However, there are often large variations in rainfall, with the year divided into wet (or monsoon) and dry seasons. This is largely due to seasonal shifts in the position of a rainy low-pressure belt called the Inter-Tropical Convergence Zone (see page 170), but in some areas is the result of monsoonal weather patterns (see page 172).

Some authorities place the start of summer and winter at their respective solstices, but these are the astronomical changes of season and often do not coincide with the meteorological changes of season. Meteorological seasons are usually determined by temperature, the onset of a particular season being defined as the date when the daily temperature reaches a certain point for a set number of days. The most common system relates to the months of the Gregorian calendar; thus, in the Northern Hemisphere, spring begins on 1 March, summer on 1 June, fall on 1 September and winter on 1 December.

Seasons can also be categorized based on ecological changes, such as the first flowering or budburst of a particular plant species. The study of the timing of ecological events is called phenology. Many ecologists recognize six seasons for temperate climate regions: prevernal (early or pre-

Below *Each year, the Earth experiences two equinoxes and two solstices. They occur when the tilt of the Earth's axis is angled parallel to, towards, and away from the Sun.*

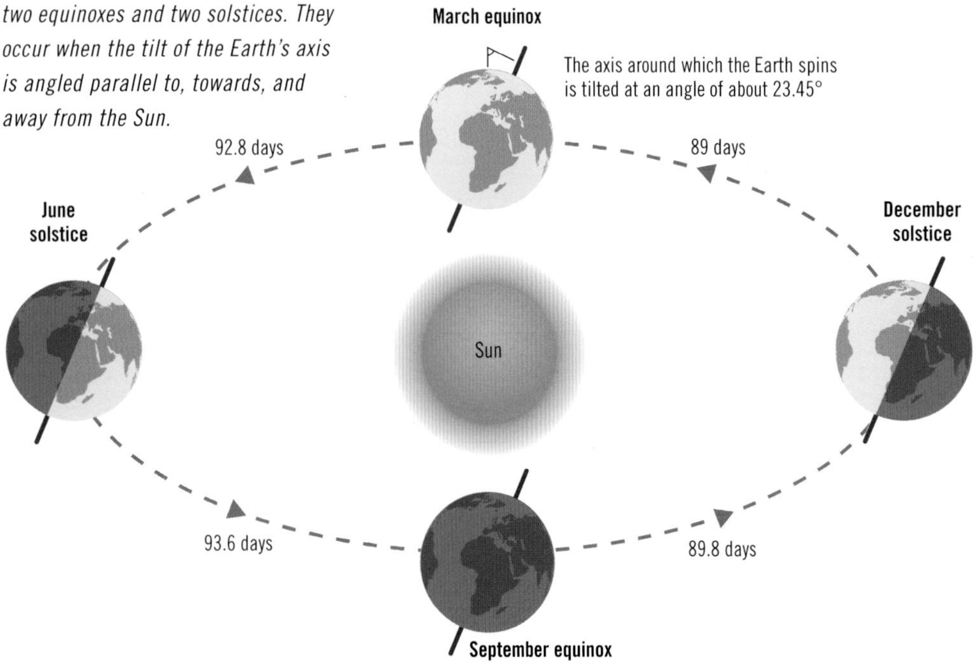

March equinox
The axis around which the Earth spins is tilted at an angle of about 23.45°
92.8 days
89 days
June solstice
December solstice
Sun
93.6 days
89.8 days
September equinox

Above *The Inter-Tropical Convergence Zone, seen here as a band of thick cloud, shifts north and south of the equator with the seasons, taking rainfall with it and thus generating the tropical wet and dry seasons.*

spring), vernal (spring), estival (high summer), serotinal (late summer), autumnal (fall) and hibernal (winter). These seasons are tied to ecological events such as budburst and bird migrations rather than fixed calendar dates, so they generally vary geographically.

The tradition of dividing the year into four seasons is a cultural construct with its origins in Western Europe. Other cultures recognize different seasonal calendars. In the Indian region, six seasons are recognized. In the Nile delta, three seasons—flood, growth, and low water—were recognized. Thailand also uses a three-season calendar (cold, hot, and rainy). Indigenous groups traditionally defined the seasons based on ecological and environmental changes, such as changing winds, flowering patterns, and animal migrations. Some North American indigenous groups use a six-season calendar, while some Australian indigenous peoples recognize eight seasons, as do the Sami people in Scandinavia.

The changes in temperature and day length that characterize seasons have a significant impact on the behavior of plants and animals, triggering periods of growth and dormancy. Many animals reproduce in spring and enter hibernation at the start of winter; many plants set fruit and then drop their leaves in fall, remain dormant over winter, flower and grow new foliage in spring, and undergo rapid growth in spring and summer.

Right *A tree in Germany in each of the four seasons.*

// Wind

The world's restless air is the result of the global patchwork of different air pressures.

Wind is generated by differences in atmospheric pressure: air flows from higher to lower pressure in order to equalize them. Ultimately, however, it is the Sun's energy that drives winds. Because the Sun heats unevenly, some areas are warmer than others, leading to the changes in air pressure that cause winds to blow.

This is illustrated by sea and land breezes. In coastal areas, by day, the land warms more quickly than the adjacent sea. The air over the land rises and cooler air rushes in from the sea to take its place, creating what is termed a sea breeze. At night, the sea holds its heat more effectively while the land becomes cooler, so the wind direction is reversed, creating a land breeze.

Wind speeds and gustiness are generally strongest during the day, when solar heating creates greater variation in air pressure. A short, high-speed burst of wind is a gust; a strong wind of intermediate duration is a squall. The strongest gust ever recorded, which reached 254 mph (408 km/h), took place on Australia's Barrow Island on April 10, 1996 during tropical cyclone Olivia. A wind's direction is usually expressed in terms of its origin, so for example, a northerly wind blows from north to south.

The Earth's rotation causes winds to be deflected by the Coriolis force everywhere except at the equator (see page 106). This means that surface winds spiral around regions of low and high pressure rather than moving directly between them. When winds reach the center of a low-pressure area, they have nowhere to go but up. As they rise, the moisture they carry then condenses into clouds, which may lead to precipitation or even storms. Similarly, at the center of areas of high pressure, dry air descends, leading to fair weather.

Winds play a significant role in determining the Earth's climate and weather, bringing moisture to the land, driving ocean currents, and steering storms. When wind patterns blowing in opposite directions collide, air is forced upwards. If one of the winds carries moisture-laden air, this can lead to a thunderstorm or tornado.

The large-scale patterns of air pressure around the globe give rise to what are known as prevailing winds. Among these are the "trade winds" and "westerlies," both generated by persistent subtropical high-pressure systems centered on about 30°N and 30°S. In the tropics, the prevailing winds blow predominantly from the northeast in the Northern Hemisphere and from the southeast in the Southern Hemisphere. These are the trade winds, which

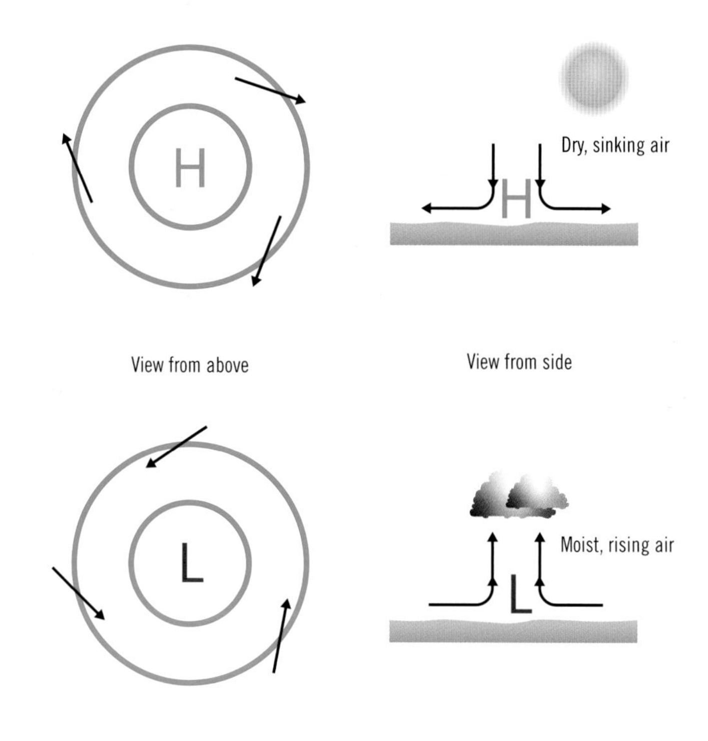

Above *Winds spiraling out from high pressure areas induce air to sink, bringing clear skies and dry conditions (top). Winds spiraling into low pressure areas cause air to rise, leading to cloudy conditions (bottom).*

steer tropical cyclones and blow African dust westward across the Atlantic Ocean. As the trade winds blow towards the equator from the north and south, they eventually converge in the Inter-Tropical Convergence Zone (see page 170), forcing air to rise and creating thunderstorms.

In the mid-latitudes (30°–60°), the prevailing winds are known as the westerlies. In the Southern Hemisphere, there is less land in the mid-latitudes, so the westerlies blow freely and can become particularly strong, peaking between 40° and 50°, where they are known as the Roaring Forties and are a key driver of the Southern Ocean circulation. They are the strongest mean sea-surface winds on Earth, reaching maximum strength during winter, when the pressure is lower over the poles. At higher latitudes (60°–90°), the prevailing winds, the polar easterlies, are often weak and irregular.

On a local scale, winds often reflect the influence of landforms—for example, sea and land breezes, mountain and valley breezes, foehn winds (warm, dry, gusty winds that flow down the leeward slopes of mountains), and katabatic winds (where cold, high-density air flows down a slope)—and local weather conditions. Rugged, mountainous topography can interrupt and distort winds by increasing friction between the atmosphere and the land or simply blocking the wind, deflecting the flow of air and creating strong updrafts, downdrafts and eddies. If forced to move through a narrow pass, winds may reach considerable speeds. Where relatively consistent winds blow across mountainous regions, the windward side of the mountains is usually moister than the leeward or downwind side.

On a smaller scale, local weather effects create shifting zones of high and low pressure that generate winds with short durations and ever-changing directions. This can create locally abrupt changes in wind speed and direction, known as turbulence.

Below *Sea breezes blow during the day, the result of warm air rising over the land and cooler air rushing in from the sea to take its place. Land breezes blow at night, the land cooling more quickly than the sea and causing the wind direction to reverse.*

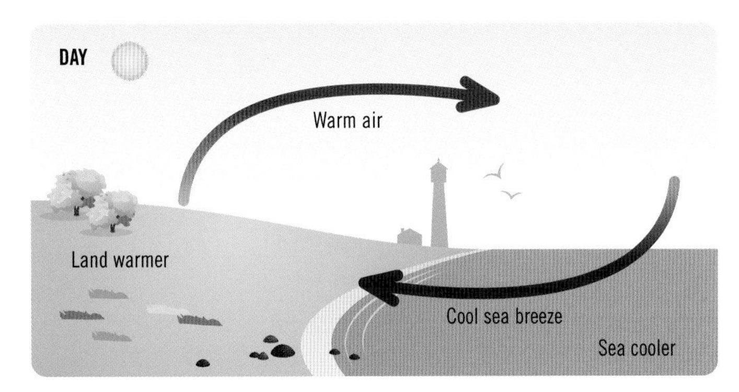

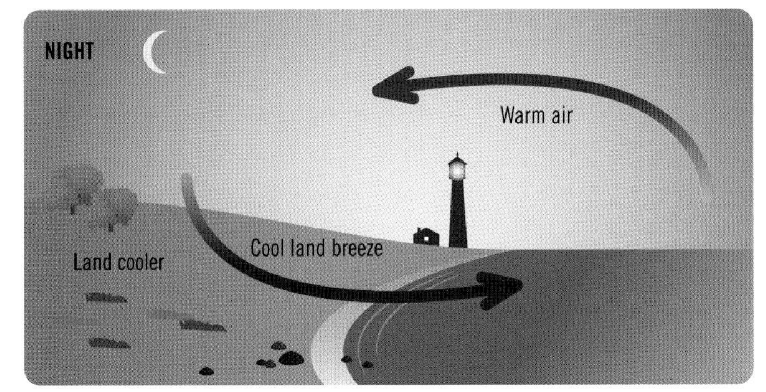

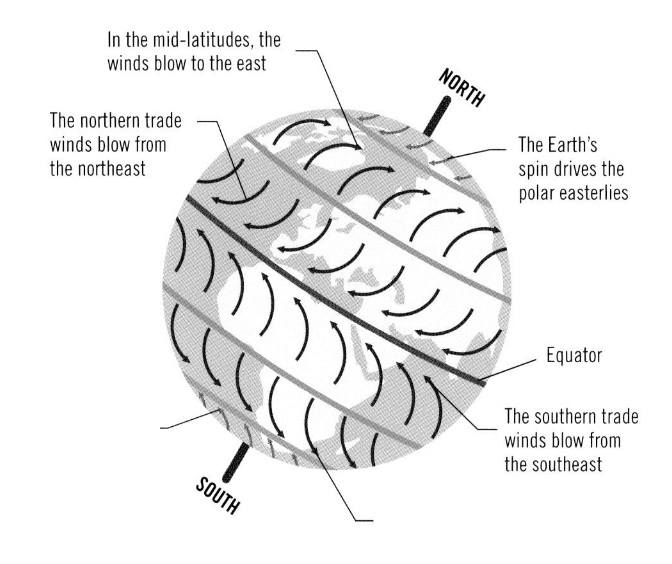

Above *The Earth's prevailing winds.*

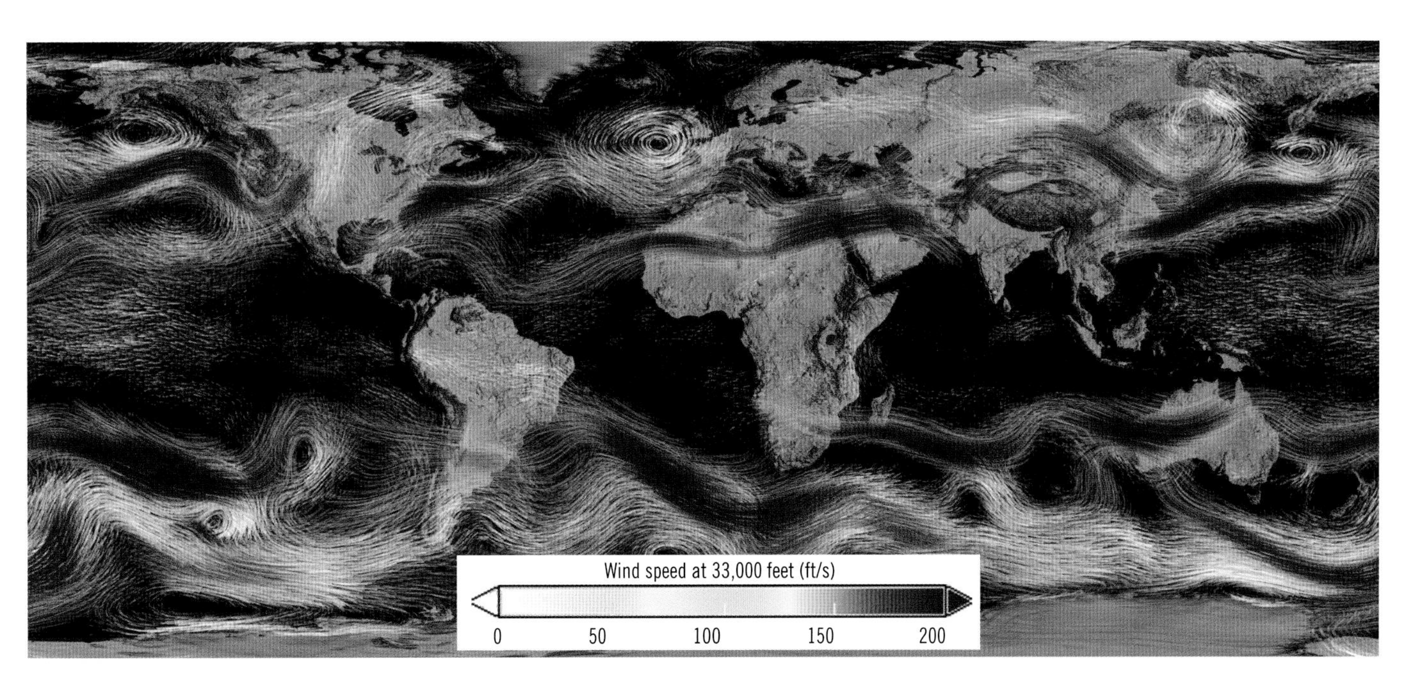

Above *The winds blowing in the upper (rainbow colored) and lower (white) levels of the Earth's atmosphere.*

// Atmospheric circulation: the three-cell model

The large-scale movement of air through the troposphere, known as atmospheric circulation, is broken up into a series of relatively self-contained cells that together distribute heat around Earth.

Atmospheric circulation is driven by the difference in the amount of sunlight—and hence heat—that falls on the equator compared to the poles. If the Earth did not rotate, the result would be a single large circulation cell in each hemisphere, with warm air rising at the equator and sinking at the poles. Instead, because the Earth does rotate, the circulation is broken up into three cells in each hemisphere—enormous rotating "donuts" of air that encircle the planet, reaching from the surface to the tropopause. The interaction of these cells with the Coriolis force is responsible for the Earth's major prevailing-wind systems. Their net effect is a transfer of energy, in the form of warm air, from the tropics to the poles, evening out the heat and moisture in the Earth's atmosphere.

Located between the equator and about 30° latitude, the Hadley cells are the largest of the three. They are driven by warm air rising over the equator. When this air reaches the tropopause, it moves towards the poles, cooling as it goes, before sinking back to the surface at around 30°N and 30°S (the so-called "horse latitudes," which are characterized by calm winds and high pressure at the surface). The dry falling air creates the zone of high atmospheric pressure that is responsible for the band of deserts that encircles the Earth at around 30°N and 30°S. As it flows back towards the equator, the air is deflected by the Coriolis force to form the easterly trade winds. The trade winds meet in a region of low atmospheric pressure known as the Inter-Tropical Convergence Zone (ITCZ), where thunderstorms are common and tropical rain forests are found. The position of the ITCZ changes with the seasons, moving north to the Tropic of Cancer (20°N) during the Northern Hemisphere summer and south to the Tropic of Capricorn (20°S) during the Northern Hemisphere winter. The region around the ITCZ was dubbed the doldrums by sailors because of the general lack of wind.

The Ferrel cells, which are found between 30° and 60° latitude, are largely driven by the movement of air in the other two cells and that caused by storms in the mid-latitudes. They flow in the opposite direction to the other two cells, with surface air blowing poleward: cold air sinks in the subtropics and rises around 60°. The Coriolis force deflects the air to form the mid-latitude westerly winds.

Below *A cross-section through the Earth's atmospheric circulation cells. The three rotating cells determine the strength and direction of the prevailing winds, as well as the position of rain belts.*

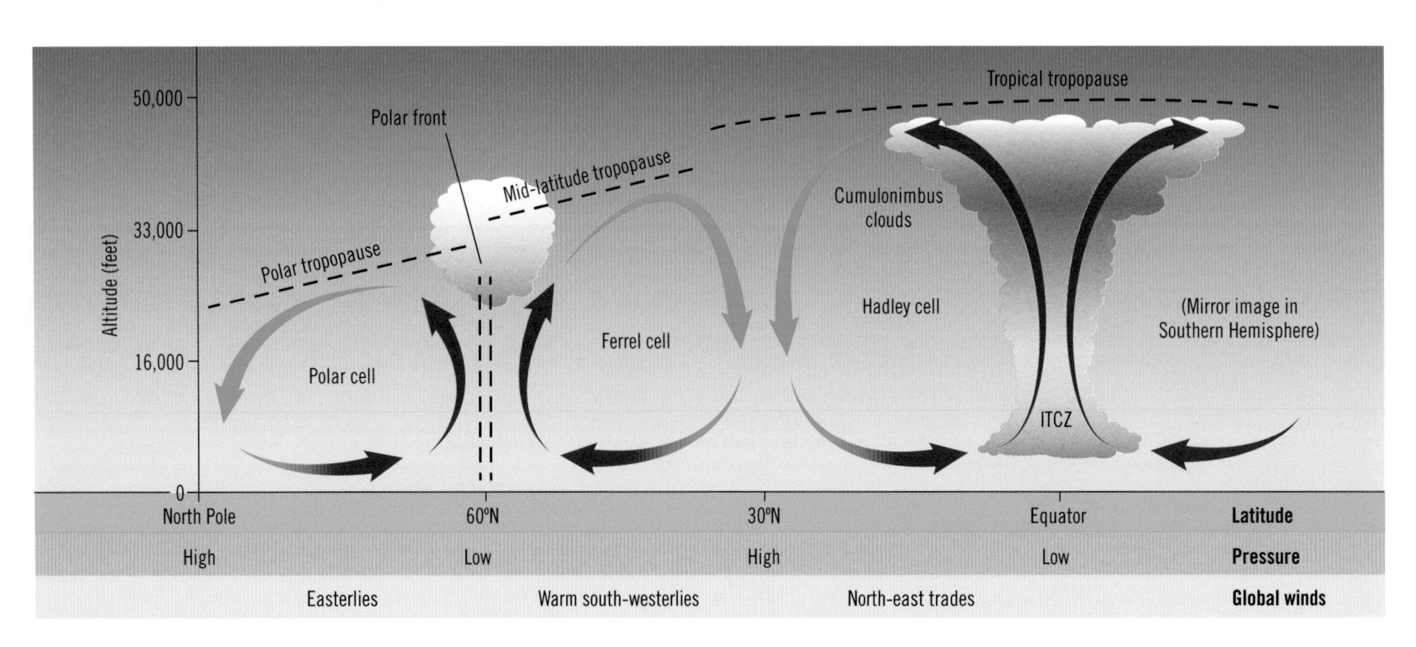

Above *A schematic view of the Earth's atmospheric circulation.*

The smallest and weakest of the three cells, the Polar cells are found between 60° and 90° latitude. They are driven by cold air sinking over the poles. This falling air creates a zone of high pressure before moving towards the equator. The air is deflected by the Coriolis force to become the polar easterly winds. When it meets the ocean or the land, the air warms up. At around 60° latitude, it rises before heading poleward once more. The boundary between the Polar and Ferrel cells, where the cold polar easterlies blowing towards the equator meet the warm subtropical westerlies blowing poleward, is known as the polar front and is associated with cloudiness and precipitation.

// The monsoon

Each year, across much of southern Asia, the weather switches abruptly from relatively dry to hot and very, very wet. This is the monsoon, a dramatic seasonal change in rainfall that is driven by a switch in the direction of the prevailing winds and the atmospheric pressure.

As described previously (see page 168), winds blow because there is an imbalance in atmospheric pressure between two locations. In the case of a monsoonal system, the pressure imbalance is the result of a disparity in the temperature of a landmass and the adjacent ocean.

The sea and the land absorb and release heat differently: land heats and cools more rapidly than water. In summer, the land surface heats quickly and becomes hotter than the adjacent ocean. This causes the air above the land to expand, creating an area of low atmospheric pressure. The air over the ocean is cooler, so is at a higher pressure, causing winds to blow from the ocean to the land and bringing moisture-laden air inland. When this air hits the land, it is driven upwards, causing it to cool and unburden itself of all that moisture, which falls as torrential rain. In a typical year, affected regions will see as much as 85 percent of their rain fall during the summer monsoon season.

In winter, the situation is reversed. The land cools more quickly than the sea, the air pressure above the land rises and the winds slowly reverse, blowing off the land and over the sea. The end of the monsoon season is more gradual than its beginning, the rainfall slowly diminishing as the land begins to cool.

Variations in the strength and timing of the Asian monsoon can have a devastating effect on the region. If the rains fail, or arrive too early or too late, it can lead to famine as the region's many small-scale farmers are unable to grow crops. Groundwater also becomes scarcer as the region's shallow aquifers are not recharged. And if

TRADE WIND

Early mariners used the regular seasonal shift in the monsoon winds to move back and forth across the Indian Ocean. Traders traveled from the Middle East to India in July–September and then, after waiting in port for the winds to change, sailed back the other way in December–February.

Below *A monsoon is an annual weather phenomenon that involves a seasonal reversal in wind direction caused by temperature differences between the land and the sea.*

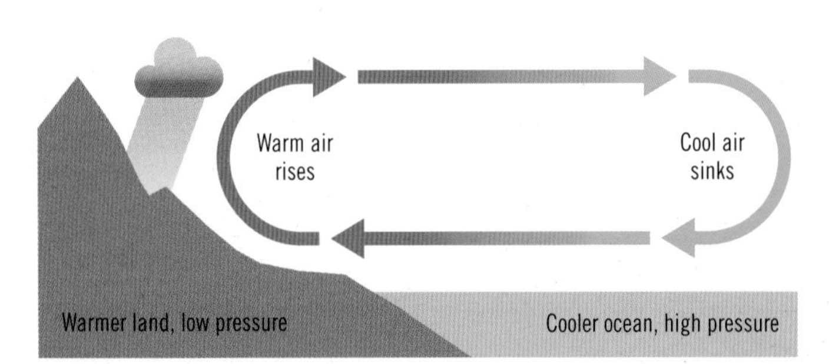

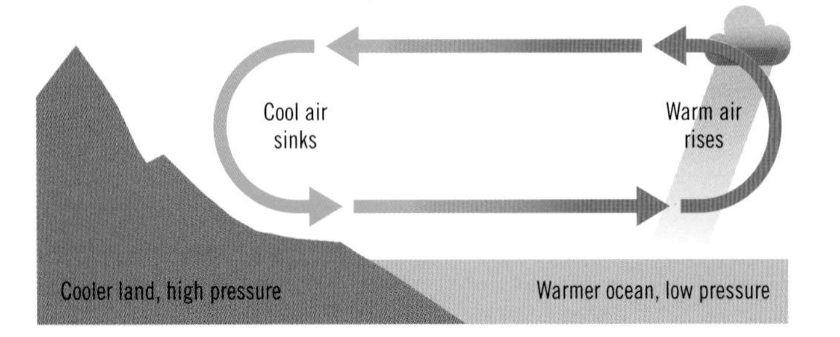

Below *Map showing the regions affected by monsoon systems.*

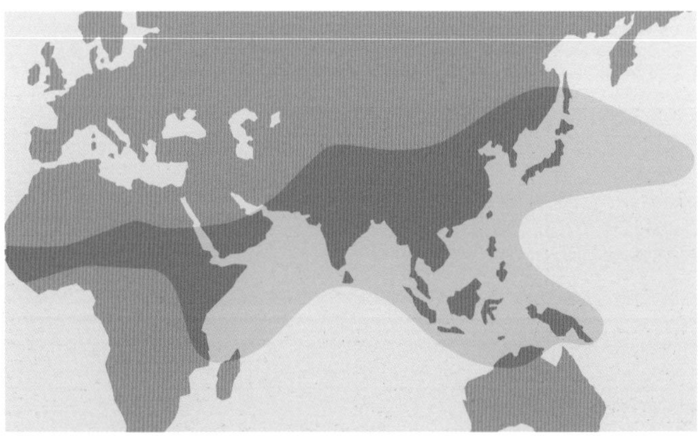

WHAT'S IN A NAME?

The word "monsoon" is derived from the Arabic word *mausim*, meaning "season." Although monsoons have both rainy and dry phases, the term is used almost exclusively to refer to the wet season.

the rains fall too heavily, it can lead to extensive flooding and powerful mudslides. In 2005, an exceptionally strong monsoon killed at least 900 people in the Indian metropolis of Mumbai; on 26 July that year, almost 3.3 feet (1 meter) of rain fell on the city.

Although Asia's annual monsoon is the best known, similar climatic conditions also occur in northern Australia, parts of western, southern and eastern Africa, and parts of North and South America.

THE WORLD'S WETTEST PLACE

The highest recorded average annual rainfall—467 inches (11,871 millimeters)—is experienced by Mawsynram, in the monsoonal northeast Indian state of Meghalaya. The most rain recorded over a single 12-month period was the 1,042 inches (26,470 millimeters) that fell on nearby Cherrapunji in 1860–61.

Monsoon rains in Mumbai, India. The monsoon typically begins in June in Mumbai, before peaking in July and tailing off in September. On average, about 95 percent of the city's rainfall takes place during this four-month period.

// Jet streams

Narrow, meandering, rapidly moving rivers of air that circulate high above the Earth, jet streams shape weather patterns around the world.

The jet streams blow in the tropopause—the boundary between the troposphere and the stratosphere (see page 130)—at a height of about 5 to 9 miles (8–15 kilometers) in both hemispheres. The stronger polar jet streams are typically located 5 to 7 (8–12 kilometers) up; the weaker subtropical jet streams occur higher—about 6–9 miles (10–15 kilometers) above sea level.

Jet streams are typically a few hundred kilometers wide and less than 3 miles (5 kilometers) deep. They blow from west to east, at speeds of around 80–140 mph (130–225 km/h), although they can reach more than 275 mph (443 km/h); the strength of the wind increases with altitude. While jet streams generally flow as a continuous stream over long distances, they often exhibit discontinuities.

A jet stream forms where cold and hot air masses meet, hence the air on the poleward side is cooler than that on the equatorward side. They are found at the transitions between the large atmospheric circulation cells (see page 170): the polar jet stream forms between latitudes 50° and 60°, where the Polar and Ferrel circulation cells meet, while the subtropical jet stream is located at around 30°, near the boundary between the Ferrel and Hadley circulation cells.

When two air masses of different temperatures meet, the difference in pressure creates wind. Ordinarily, the wind would flow from the hot to the cold air mass, but the Coriolis effect causes it to be deflected and it instead flows along the boundary of the two air masses.

The positions of the jet streams are not fixed and they regularly shift to the north or south as they follow the boundary between the hot and cold air masses. They may also dip or rise in altitude. In consequence, they typically take a sinuous, meandering path and may split into two or more parts, stop and restart elsewhere, and form eddies. The meanders, which propagate towards the east at lower speeds than the flow itself, are known as Rossby waves.

The jet streams also change strength and position with the seasons. The greater temperature differential between the tropics and the poles during winter makes them blow more strongly. The polar jet streams are said to follow the Sun: as the Sun's elevation increases during spring, the jet streams' average latitude shifts towards the poles; during fall, they begin to head equatorward.

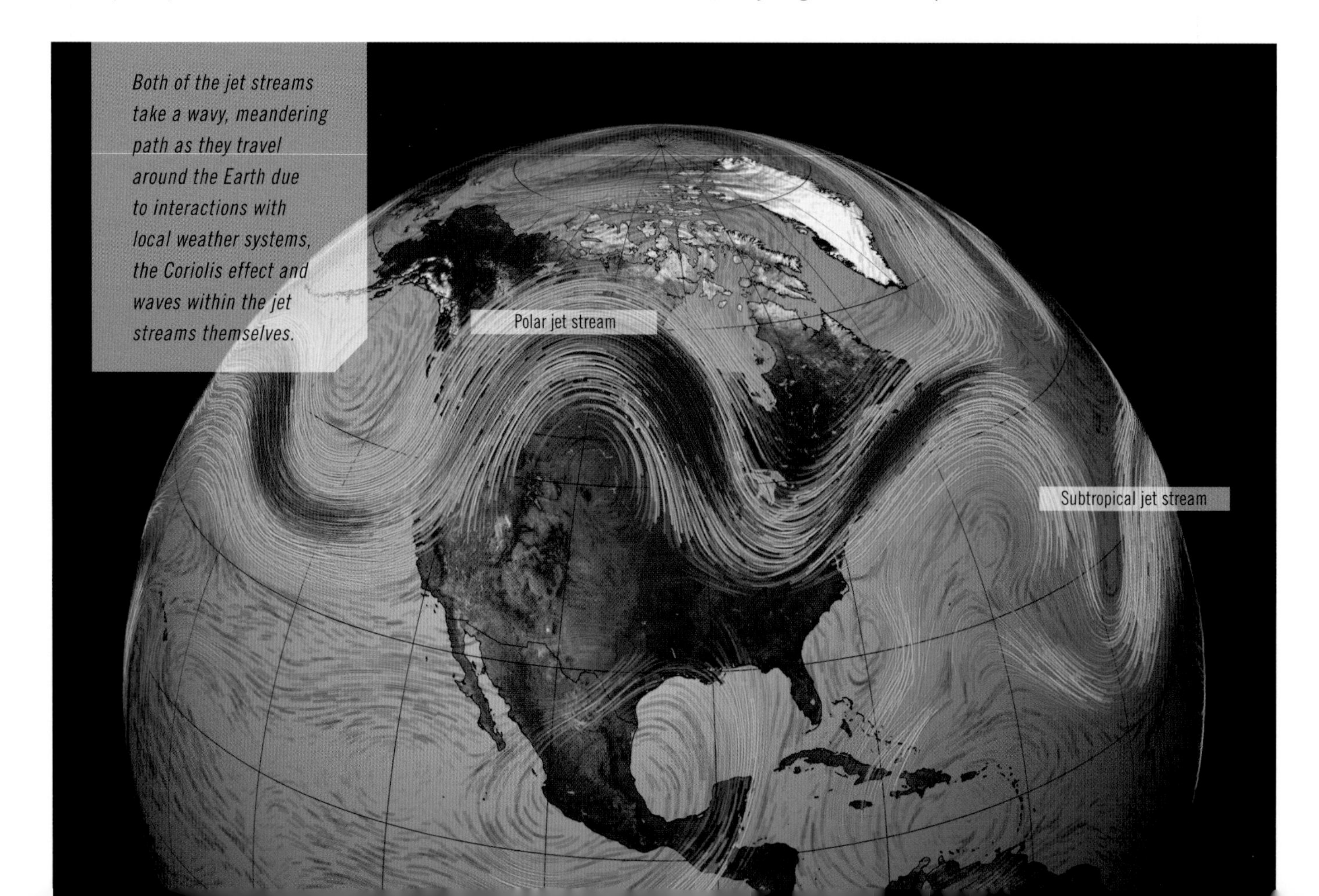

Both of the jet streams take a wavy, meandering path as they travel around the Earth due to interactions with local weather systems, the Coriolis effect and waves within the jet streams themselves.

The El Niño-Southern Oscillation (see page 164) is another factor that causes the upper-level jet streams to shift their positions and change strength, leading to changes in precipitation, temperature, and storm development across a broad geographical area. During an El Niño, the northern polar jet stream strengthens and moves south, and the northern subtropical jet stream also strengthens. Meanwhile, in the Southern Hemisphere, the subtropical jet stream is displaced north of its normal position.

When a jet stream changes position, it can have a significant impact on the weather, so meteorologists use their locations as a forecasting aid.

Jet streams are important for air travel because of their impact on fuel use. Pilots will often attempt to fly close to and with the flow of a jet stream, gaining a boost from the strength of the tailwind. However, doing so can increase the likelihood of the plane encountering potentially dangerous clear air turbulence.

Below *The stronger polar jet stream is typically located 5 to 7 miles (8–12 kilometers) up between latitudes 50° and 60°, where the Polar and Ferrel circulation cells meet, while the weaker subtropical jet stream is located about 6–9 miles (10–15 kilometers) up at around 30°, near the boundary between the Ferrel and Hadley circulation cells.*

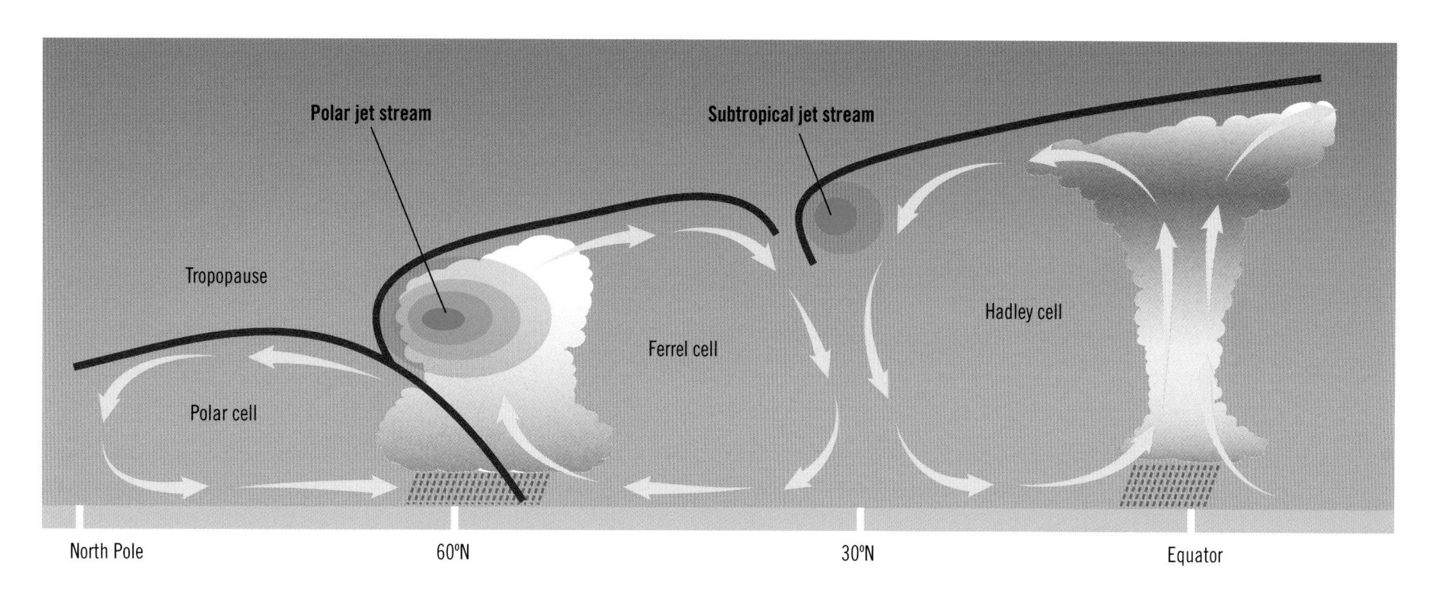

POLAR-NIGHT JET STREAM

From time to time during winter, a jet stream develops within the stratosphere at an altitude of about 16 miles (25 kilometers) and a latitude of roughly 60°. Known as the polar night jet, it typically breaks down again in either late winter in the Northern Hemisphere or mid- to late spring in the Southern Hemisphere.

Right *El Niño events cause the northern polar jet stream to strengthen and shift southwards, while the northern subtropical jet stream also strengthens, bringing increased rainfall to the southern USA and Mexico.*

The polar vortex

High above the North and South Poles sit large, rotating systems of low pressure that periodically bring extreme cold weather to the mid-latitudes.

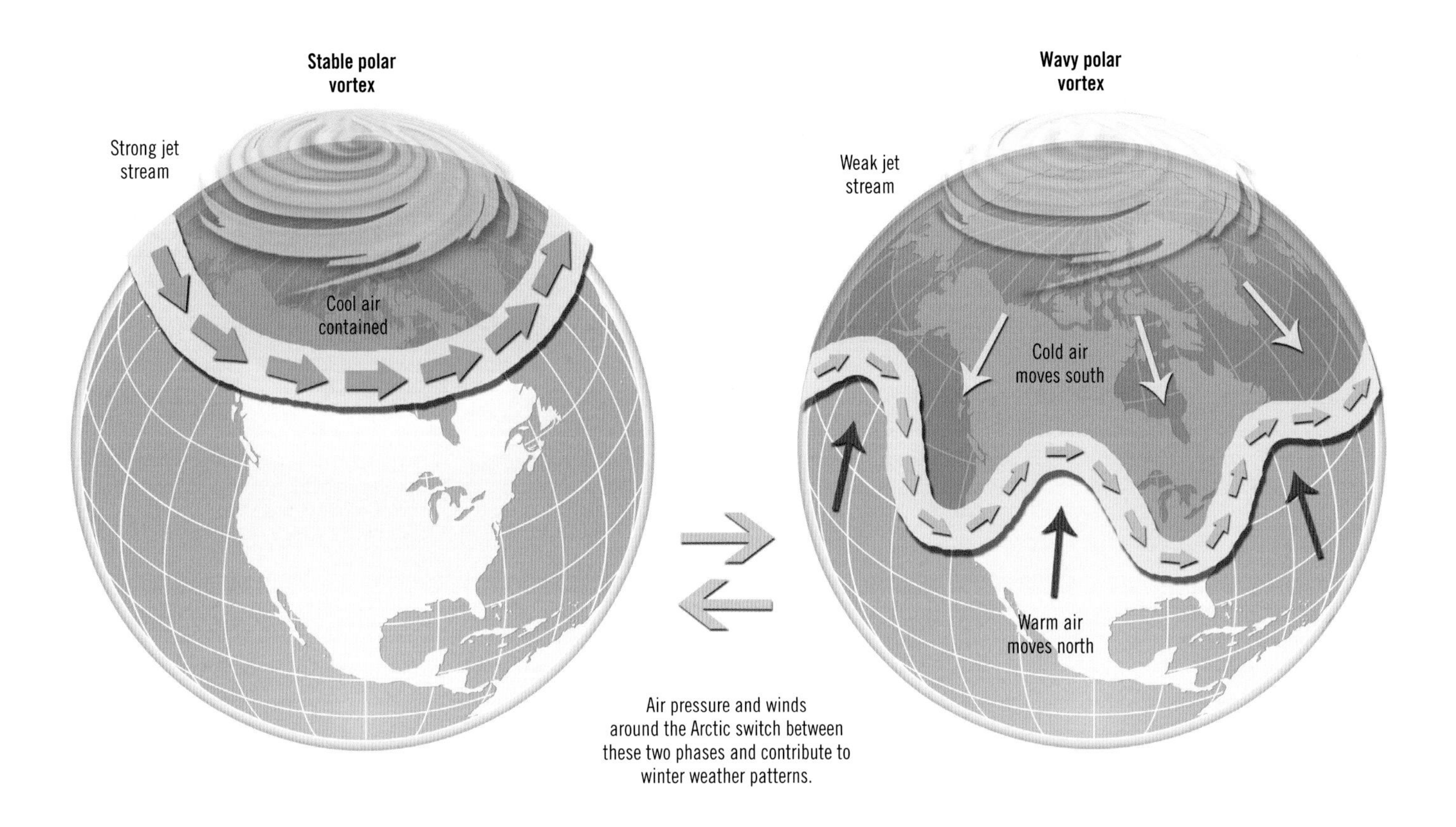

Above *A stable polar vortex keeps the jet stream blowing in a near-circular route, but if it weakens, the jet stream will meander more.*

Over each of the Earth's poles are two regions of low pressure—one in the troposphere and one in the stratosphere—known as polar vortexes. The two are not directly connected, but very occasionally they do interact. As with other cyclonic systems, the Coriolis effect causes the vortexes to rotate in a counter-clockwise direction in the Northern Hemisphere and a clockwise direction in the Southern Hemisphere. Both the tropospheric and stratospheric Southern Hemisphere polar vortexes tend to be more uniform and more stable than their Northern Hemisphere counterparts.

The tropospheric polar vortexes are present year-round, although they weaken in summer and strengthen in winter. They are driven by the temperature difference between the Arctic or Antarctic and the mid-latitudes and are bounded by the polar jet streams. These are by far the larger of the two types of polar vortex: the edge of the vortex generally lies between latitudes 40° and 50°, compared to about 60° for the stratospheric polar vortex.

Under normal conditions, this swirling atmospheric phenomenon keeps a mass of frigid air confined over the pole, but during winter it often expands, sending cold air into the lower latitudes. Pockets of frigid air can even break off and migrate away from the polar regions, bringing sub-zero temperatures to the mid-latitudes. The vortex can also contract, bringing warmer weather to the high latitudes.

The stratospheric polar vortexes form each autumn, strengthen over winter, and then dissipate in spring. They are roughly circular and are bounded by the polar night jets. Like the tropospheric polar vortexes, they form as a consequence of the large-scale temperature gradients between the mid-latitudes and the pole.

Both types of vortex can play a role in extreme weather events at the surface, but the tropospheric vortex is most often involved. A strong tropospheric vortex keeps the polar jet stream blowing around the Earth in a broadly circular path centered on the pole. However, when the vortex is weakened, the low-pressure system exerts less control over the jet stream and it begins to follow a more wavy, rambling path, with pockets of cold air moving away from the polar regions and warmer air moving towards them.

In January 2019, a large mass of air escaped from the Arctic tropospheric polar vortex and moved into Canada and the US Midwest, bringing record low temperatures (on one particular day, Chicago was colder than Antarctica) and heavy snowfalls to many areas. At least 22 people died in the cold snap before it finally ended in March.

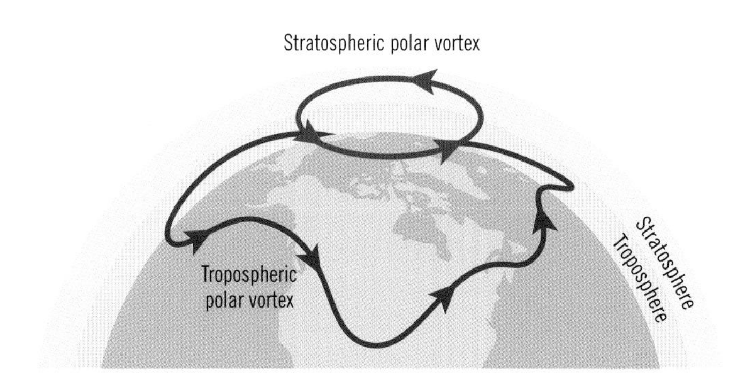

Above *The stratospheric polar vortex sits above the tropospheric polar vortex, but sometimes air within the former drops and affects the latter.*

Sudden stratospheric warming

Occasionally (about six times a decade), a disturbance in the troposphere causes the Northern Hemisphere stratospheric polar vortex to break down. When planetary-scale atmospheric waves (known as Rossby waves) from the troposphere travel up into the stratosphere and break, they cause the westerly winds of the polar vortex and polar night jet to rapidly decelerate or even reverse direction. The vortex may then split into two smaller vortexes or be pushed from its usual position, centered near the pole, to being centered over northern Siberia.

These are known as sudden stratospheric warming (SSW) events because they can cause temperatures in the stratosphere to warm by as much as 90°F (50°C) in just a few days. SSW events are some of the most extreme atmospheric phenomena. (A major Southern Hemisphere SSW event has been observed only once.)

As the polar night jet's winds slow, they begin to turn towards the center of the polar vortex. When the air reaches the center, it is forced to descend, which causes it to compress. This, in turn, causes the air's temperature to rise dramatically and the air pressure above the North Pole to increase.

About a third of SSW events have little detectable effect at the surface, but if the cold descending stratospheric air reaches the troposphere, it can cause the polar jet stream to meander and bring cold air to the mid-latitudes and warm air to the Arctic. It can take from a few days to a few weeks for an SSW event to influence weather at the surface.

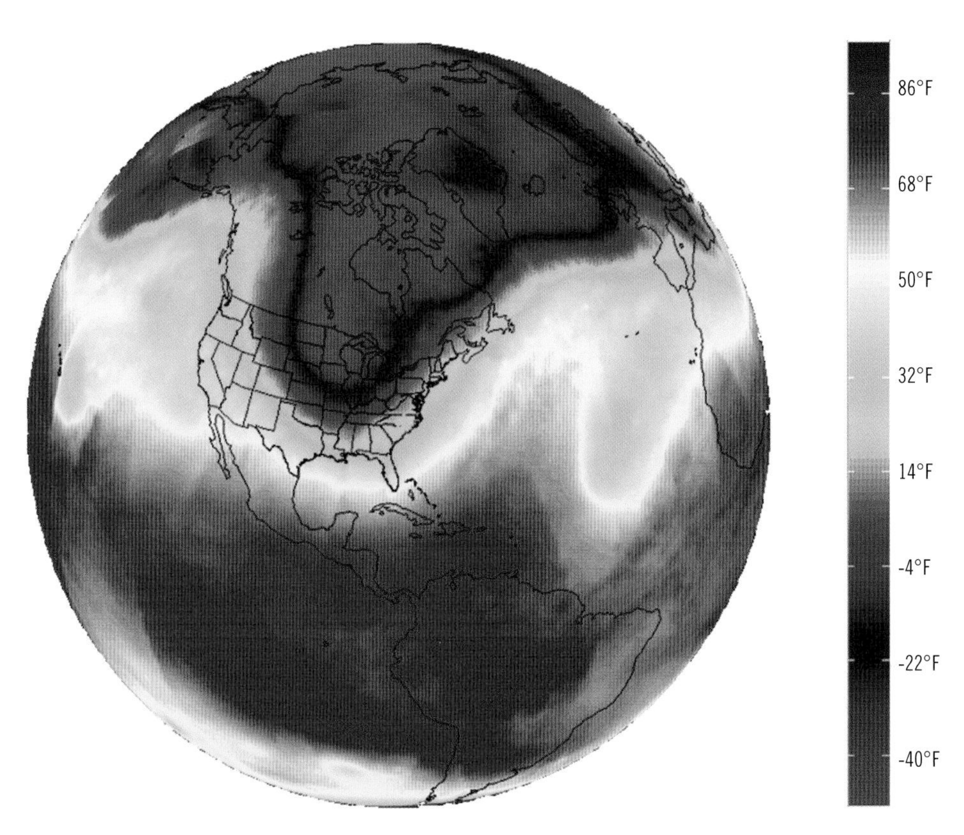

Left *This NASA satellite image shows the result of a sudden stratospheric warming event in January 2019 that caused a mass of frigid air to dip down into Canada and the US Midwest, killing at least 22 people. Parts of Minnesota experienced temperatures as low as -47°F (-44°C).*

// The geomagnetic field

Essentially an enormous bar magnet, the Earth produces a magnetic field that is vital for the survival of life on the planet.

Magnetic field strength on the Earth's surface in nanoteslas.

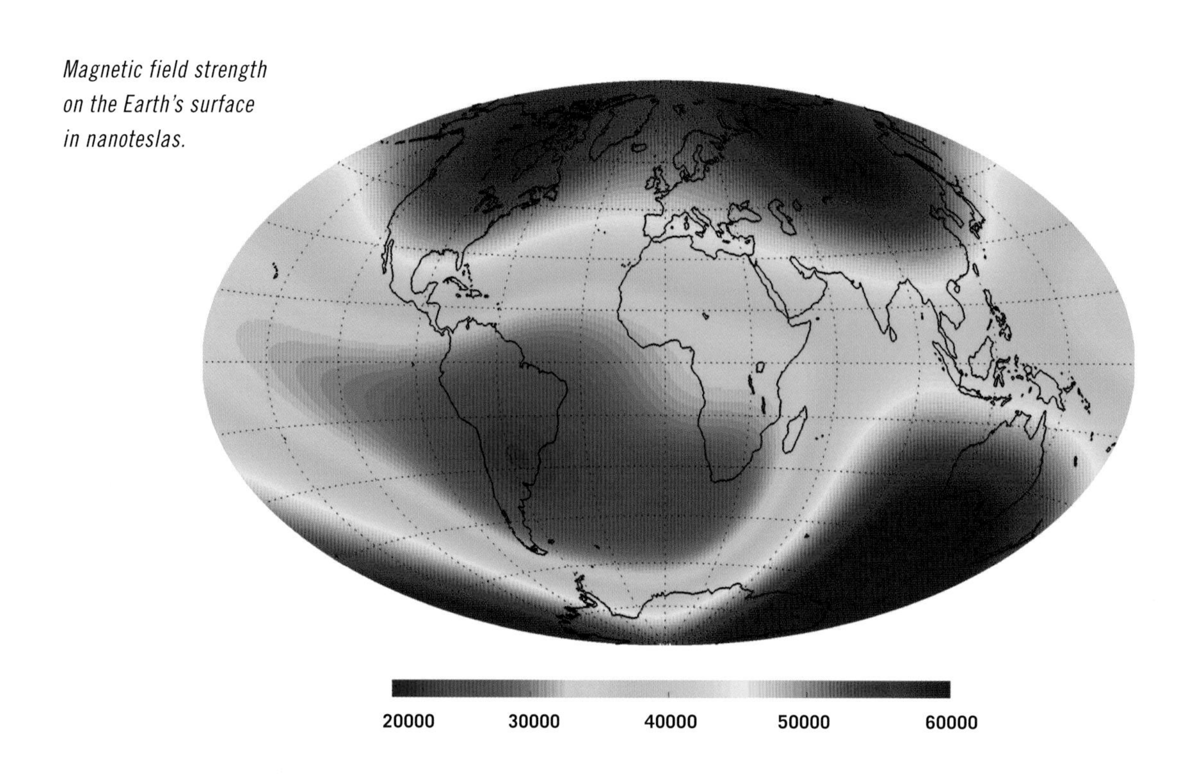

The Earth's magnetic field, or geomagnetic field, is generated by what is known as a dynamo effect. Heat escaping from the Earth's solid core creates convection currents within the molten iron and nickel in the outer core, which in turn generate electric currents. As the Earth rotates on its axis, these electric currents, which are aligned in a north–south direction, create a magnetic field that extends around the planet. The Earth's magnetic field is believed to be at least 3.45 billion years old.

In the outer core, the strength of the geomagnetic field is about 500 microteslas. At the Earth's surface, it ranges from about 25 to 65 microteslas—about 100 times weaker than a typical refrigerator magnet. The intensity tends to decrease from the poles to the equator. Over the past 150 years, the strength of the overall geomagnetic field has decreased by just over 10 percent.

The Earth's magnetic axis, its dipole, is inclined at an angle of about 10° to the axis around which the planet spins; in other words, the geographical and magnetic poles do not line up. (There is also a slight difference between what are known as the geomagnetic poles, located where they explain observed global magnetic patterns, and the actual magnetic poles that a magnet points towards.)

Although the Earth's geomagnetic poles are usually located near the geographical poles, they are continuously changing position, albeit very slowly. The two magnetic poles meander independently of one another. At present, the magnetic North Pole is drifting to the northwest, from northern Canada towards Siberia. The rate at which it is moving is accelerating, from 6.2 miles (10 kilometers) per year at the beginning of the twentieth century to 25 miles (40 kilometers) per year in 2003.

Every so often, the North and South Geomagnetic Poles trade places. The history of these reversals can be read on the sea floor. As new sea floor is created at mid-ocean ridges, the cooling magma records the direction of the geomagnetic field in strongly magnetic minerals, particularly iron oxides such as magnetite (see page 102). Reversals appear to occur relatively quickly and effectively at random; intervals between events range from less than 100,000 years to as much as 50 million years, with an average interval of about 300,000 years. The most recent reversal took place about 780,000 years ago. It is unclear why these reversals take place.

The geomagnetic field is what makes a compass usable

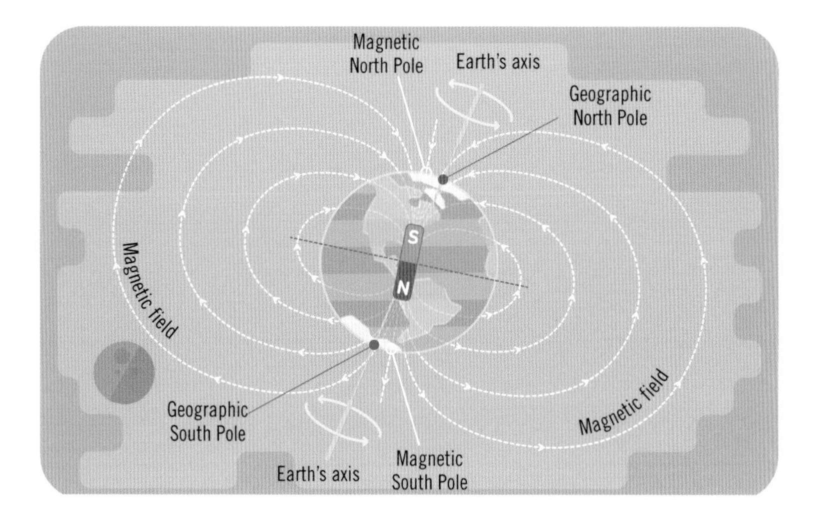

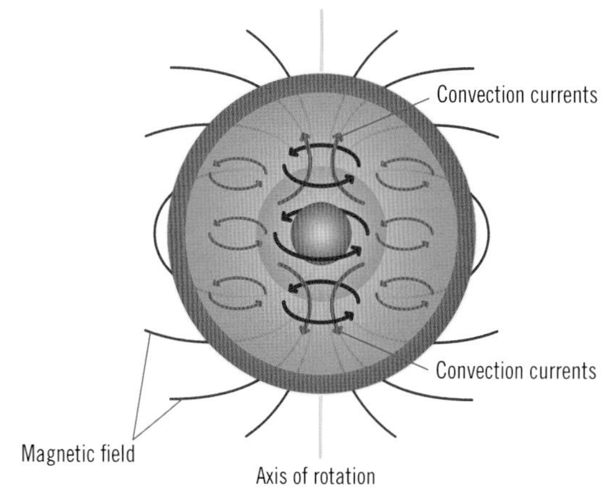

Above *Acting like a huge bar magnet, the Earth's magnetic field is generated by convection currents within the planet's core.*

for navigation. Some animals can detect the geomagnetic field and use it to navigate during their migrations.

The magnetosphere

The region of space around the Earth in which the dominant magnetic field is that produced by the Earth itself is known as the magnetosphere. Extending several tens of thousands of kilometers into space, its shape is constantly changing as it is buffeted by the solar wind, a continuous flow of plasma from the Sun made up mostly of electrons and protons. The solar wind compresses the magnetosphere on the side facing the Sun and stretches it into a long tail on the far side. The boundary between the solar wind and the magnetosphere is called the magnetopause. By deflecting the solar wind around the Earth, the magnetosphere protects life on Earth, which would otherwise be exposed to high levels of radiation. It also helps to keep the Earth's atmosphere intact; without it, the solar wind and cosmic rays would strip away the upper atmosphere, including the ozone layer.

During periods when the solar wind is particularly strong, charged particles flow down the magnetic field lines at the poles and strike gas atoms in the atmosphere, causing auroras.

Below *The Earth's magnetosphere protects it from the solar wind, a continuous flow of plasma from the Sun made up mostly of electrons and protons. The impact of the solar wind compresses the magnetosphere on the side facing the Sun and stretches it into a long tail on the far side.*

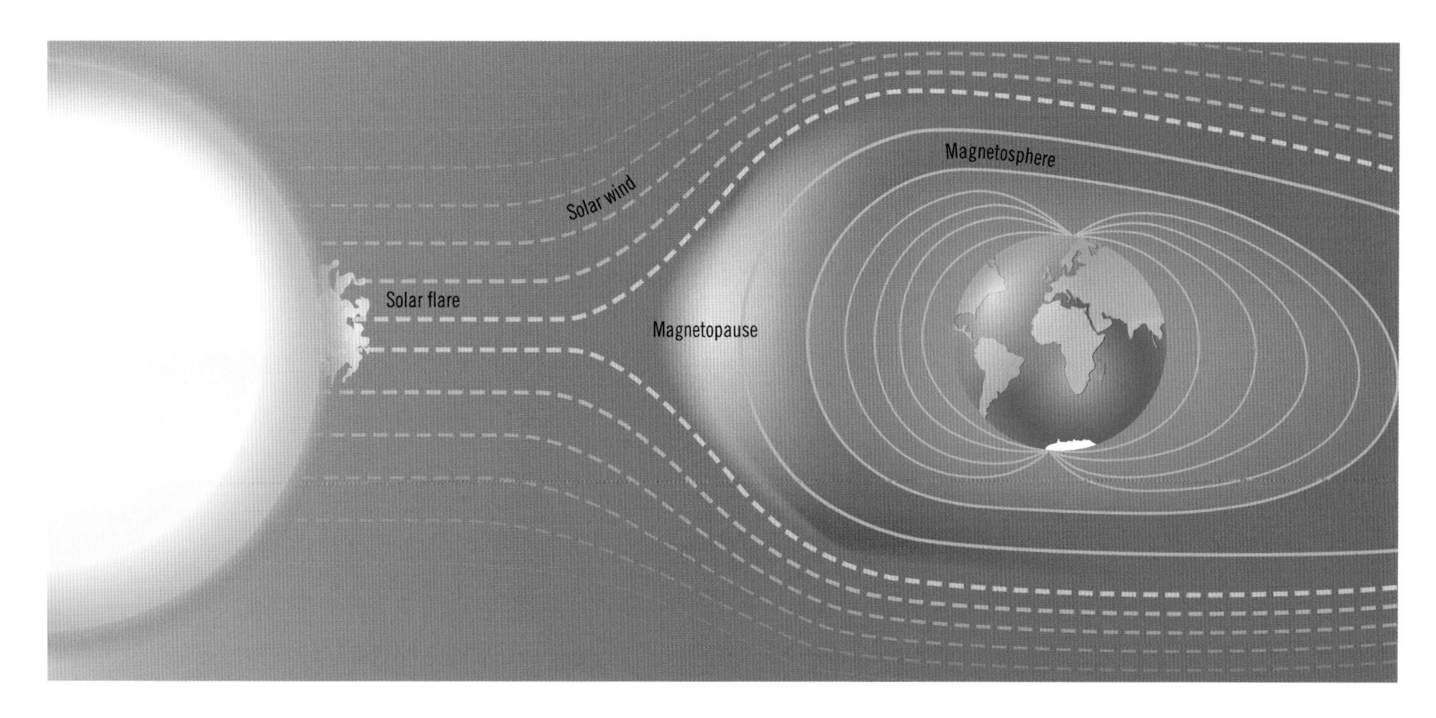

// The auroras

The skies over the high latitudes are periodically lit up by spectacular, colorful light displays—the auroras.

In the Northern Hemisphere, the aurora is known as the Aurora Borealis or the Northern Lights; in the Southern Hemisphere, it is called the Aurora Australis or the Southern Lights.

Auroras are caused by collisions between charged particles (mostly electrons but also protons) that have been released from the Sun and atoms and molecules of gas in the Earth's upper atmosphere. The collisions cause the gas atoms and molecules to be either "excited" or ionized. They then release a photon of light when they return to their normal state. Different gas atoms and molecules produce different-colored light: oxygen produces green light at low altitudes (the most common auroral color) and red light at high altitudes, while nitrogen glows blue and purple.

Although the Sun releases a constant stream of charged particles—a rarefied flow of hot, magnetized plasma known as the solar wind—this is not enough to cause an aurora. Most of the time, electrons and protons in the solar wind are deflected away from the Earth by the Earth's magnetic field (see page 178). However, the Sun periodically experiences "storms" that can intensify the solar wind. The most powerful solar storms are coronal mass ejections, when the Sun's corona releases a significant burst of plasma and an accompanying magnetic field. If they occur on the part of the Sun that faces the Earth, these solar storms will trigger an aurora.

Above *The Aurora Borealis over Jökulsárlón Glacier Lagoon, Iceland.*

Below *The Aurora Australis as seen from the International Space Station.*

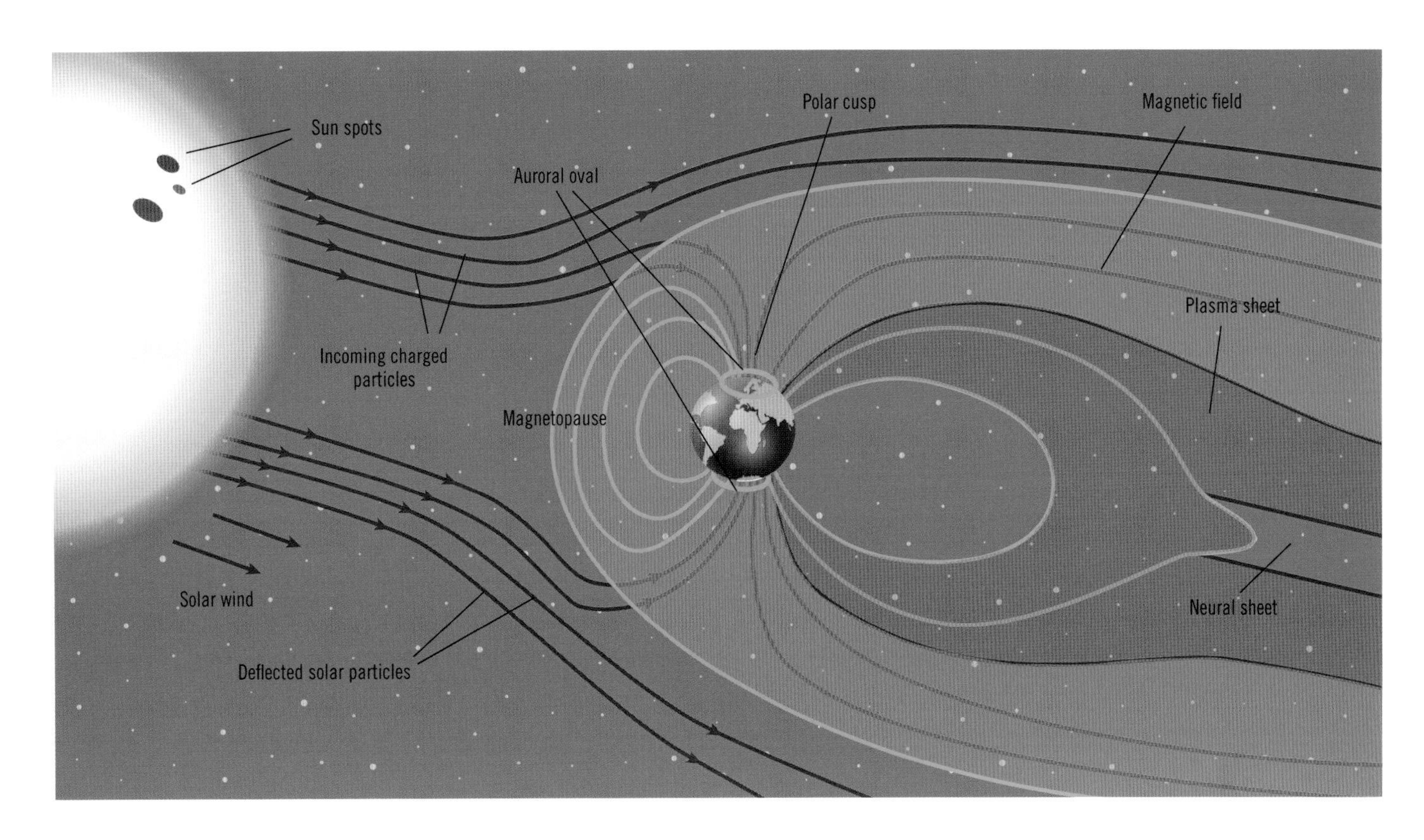

Above *Most of the time, the solar wind is deflected by the Earth's magnetic field, but when solar flares occur, the high-energy plasma is funneled down into the polar regions, where it interacts with gas atoms and molecules, creating the auroras.*

The Sun goes through an 11-year cycle of activity; during its most intense phase, coronal mass ejections become more frequent and significantly increase the intensity of the solar wind. During these periods, auroras become both more frequent and brighter.

Auroras take several different forms: a mild glow, often near the horizon; arcs that curve across the sky from horizon to horizon like a rainbow; ribbon-like bands; patches or surfaces that look like clouds or curtains; pillars or rays that appear as columns, filaments, or streaks; and coronas that diverge out from a single point in the sky to cover a large area. The form that an aurora takes is partially dependent upon the amount of acceleration that is imparted to the charged particles by the geomagnetic field. When an aurora intensifies, the patterns can begin to "dance," shifting and flowing quite rapidly. In most instances, the northern and southern auroras mirror each other, changing simultaneously with similar shapes and colors. The light from a particularly spectacular display can be so bright that it is possible to read a newspaper at night.

The auroras typically take place between about 50 and 93 miles (80–150 kilometers) above the Earth's surface, but can be as high as 400 miles (640 kilometers) up. The geographical region in which an aurora takes place is called the auroral oval. It is usually found in a band known as the auroral zone, a region about 3° to 6° of latitude wide located between 10° and 20° from the geomagnetic poles. An increase in solar activity causes the two auroral ovals to expand, so that auroras occur at lower latitudes.

Auroras sometimes create a hissing or crackling noise or even muffled bangs, beginning about 230 feet (70 meters) up. These noises are thought to be created when charged particles interact with an inversion layer in the atmosphere.

Below *This map shows the regions in which the auroras most often occur.*

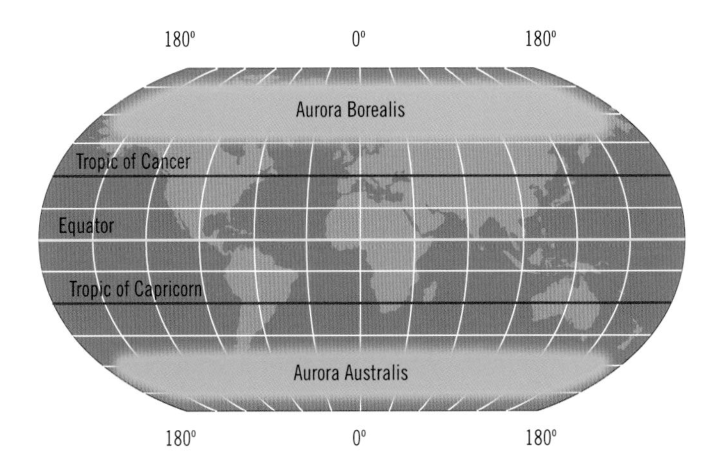

// Climate change

The greenhouse effect

The rise in the concentration of atmospheric gases that trap heat has caused the Earth's climate to change.

The greenhouse effect describes the warming that takes place when certain gases in the Earth's atmosphere (mostly carbon dioxide, methane, nitrous oxide, and water vapor—collectively known as greenhouse gases) trap heat. Just like the glass in a greenhouse, these gases let sunlight in, but then stop the heat that it creates from escaping.

In its natural form, the greenhouse effect is responsible for keeping the Earth habitable. Without it, the planet's surface would be an average of about 59°F (33°C) cooler.

During the Earth's long history, levels of greenhouse gases have risen and fallen, and the global temperature has followed suit. For the past several thousand years, however, greenhouse gas concentrations and temperatures have both remained relatively stable—until recently.

Below *The greenhouse effect. Incoming solar radiation is either reflected back into space or absorbed by the land, sea and atmosphere, causing them to heat up. Some of this heat is then radiated back into space, but some of it is absorbed by greenhouse gases in the atmosphere. The higher the concentration of greenhouse gases, the more heat is retained.*

The burning of so-called fossil fuels (hydrocarbons such as oil, gas, and coal formed from the remains of dead plants and animals; see page 60) and other activities have resulted in the emission of large amounts of greenhouse gases, enhancing the greenhouse effect and significantly warming the Earth. Since the pre-industrial period (which ended around 1850, although there is no specific date and a shortage of actual measurements from this time), human activity has raised atmospheric carbon dioxide levels by almost half. Atmospheric concentrations of carbon dioxide, methane and nitrous oxides are the highest they have been for some 800,000 years (possibly as long as five million years for carbon dioxide).

As in the distant past, the recent rise in the concentration of greenhouse gases has resulted in an increase in global surface temperatures. Since the pre-industrial period, the Earth's global average temperature has risen by about 1.8°F (1°C) and is continuing to rise by 0.4°F (0.2°C) every ten years. All but one of the 16 hottest years on record have occurred since 2000. Predictions

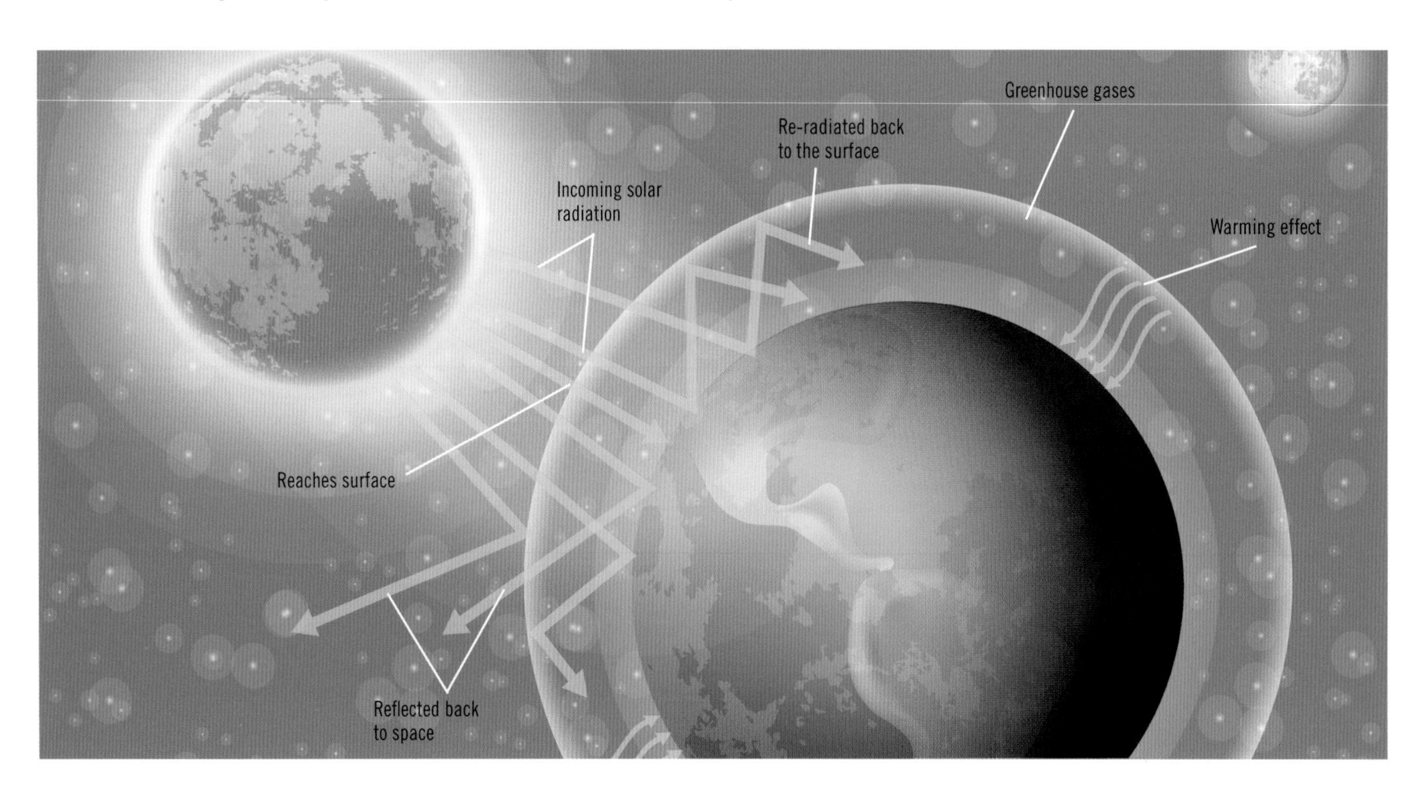

Above *Industrial activity is responsible for about a quarter of global greenhouse gas emissions, mostly released through the burning of fossil fuels for energy, but also from the chemical reactions necessary to produce goods from raw materials.*

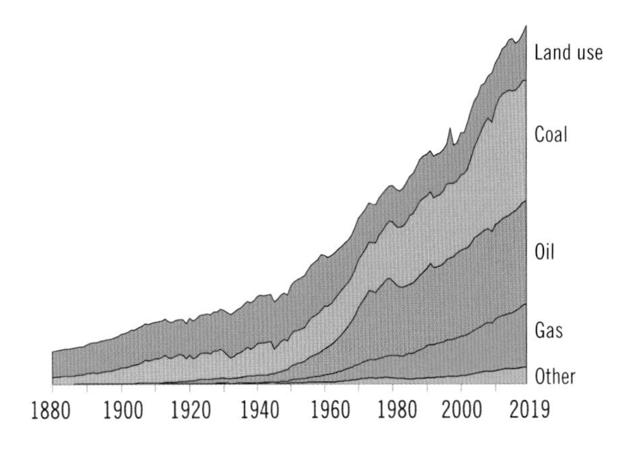

Above *A breakdown of global carbon dioxide emissions since 1880 by source shows the overwhelming dominance of fossil fuels.*

suggest that if carbon emissions continue at their current rate, by 2100 the global mean surface temperature will have increased by 5.4–7.2°F (3–4°C) compared to the 1986–2005 average.

Water vapor is the most potent greenhouse gas in the Earth's atmosphere; however, its impact on warming is complicated. Warming leads to increased evaporation, resulting in more water vapor in the air, so more warming. But increasing water vapor also leads to more clouds, which means more sunlight is reflected, so less warming.

Carbon dioxide is the most significant greenhouse gas. It is mainly produced by the burning of fossil fuels, forest clearing and burning, and cement production. Each year, about 40 billion tons (36 billion tonnes) of carbon dioxide are released through human activities. About half of this total is absorbed by natural sinks (see box), mostly the oceans, but the remainder stays in the atmosphere.

Methane is a more potent greenhouse gas than carbon dioxide. Over short timescales, it traps about 85 times more heat per molecule, but its atmospheric concentration is far lower and its residence time is much shorter (about ten years, compared with hundreds of years for carbon dioxide), so its impact is not as great. Major sources of methane include rice cultivation, livestock farming, the burning of coal and natural gas, and the decomposition of organic matter in landfill. Levels of methane in the atmosphere are now almost three times pre-industrial levels.

SINKS AND SOURCES

Carbon sinks are reservoirs in which carbon is absorbed and stored through a process known as carbon sequestration. Examples include forests and oceans. Carbon sources are sites and processes that produce chemical compounds composed of carbon, such as carbon dioxide and methane. Examples include farms and combustion engines. Put simply, carbon sinks absorb more carbon than they release, while carbon sources release more carbon than they absorb.

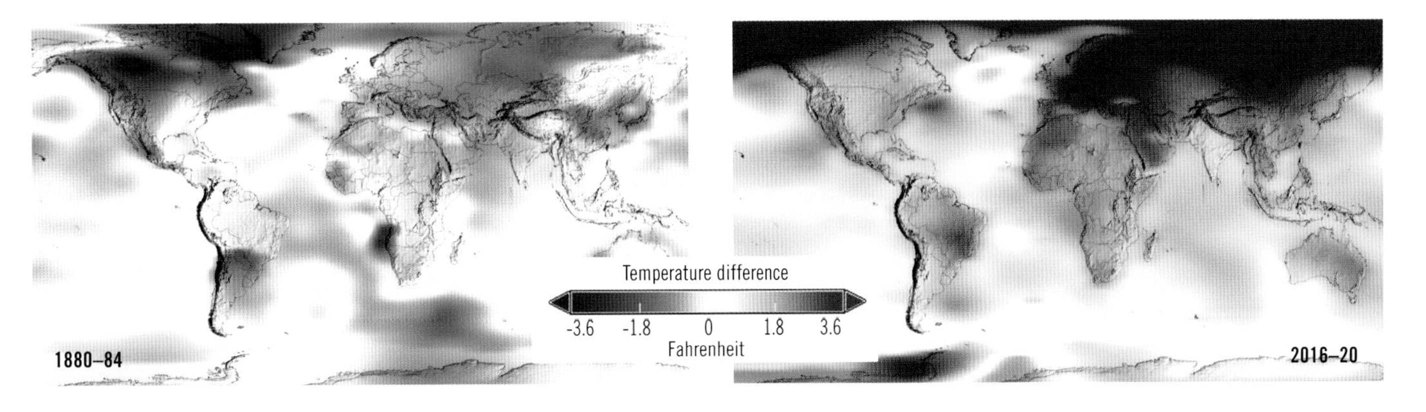

Above *These two images show the global temperature anomalies (the amount that temperatures differ from the average over the 30-year baseline period 1951 to 1980) for 1880 to 1884 (left) and 2016 to 2020 (right). Higher than normal temperatures are shown in yellow and red, while lower than normal temperatures are shown in blue. Extreme warming in the Arctic can be seen at the top of the right image.*

Feedbacks and tipping points

As the world warms, there are fears that higher temperatures will trigger climate feedbacks which will accelerate the warming process.

Feedback loops involve a change that creates conditions that have an impact on the direction of the change, either reinforcing it or diminishing it. Now that global warming is underway, there are several such phenomena, known as climate feedbacks, with the potential to either accelerate (positive) or slow (negative) that warming.

Perhaps the most worrying positive feedbacks are to be found in the Arctic. Over the past 30 years, the Arctic has warmed about twice as rapidly as the mid-latitudes, a phenomenon known as Arctic amplification.

Melting of the Arctic sea ice reduces the overall albedo (reflectivity) of the region, meaning that more of the Sun's heat is absorbed, particularly by sea water, which leads to more melting (the loss of sea ice is a key driver of Arctic amplification). Similarly, rising temperatures are causing the Arctic permafrost to thaw, resulting in the release of large quantities of stored carbon, which raises temperatures further and causes more permafrost to thaw (estimates suggest that thawing Arctic permafrost is already releasing an estimated 1.9 billion tons/1.7 billion tonnes of carbon annually). Drier conditions are leading to more frequent wildfires, which are also releasing huge amounts of carbon. And finally, warming waters are increasing the risk that gas hydrate deposits will melt, causing the sudden release of vast amounts of methane (global gas hydrate deposits are thought to contain about twice the carbon contained in all coal, oil, and conventional natural gas reserves combined), which would lead to further warming.

Globally, rising temperatures risk turning current carbon sinks (see page 182) into carbon sources. For example, the Earth's soils contain about 2,500 gigatons (2,300 gigatons) of carbon—more than three times the amount in the atmosphere—but as the world warms, soils are increasingly releasing that carbon back into the atmosphere. Similarly,

Below *This diagram shows how two important climate feedbacks—evaporation and melting ice—act to accelerate global warming.*

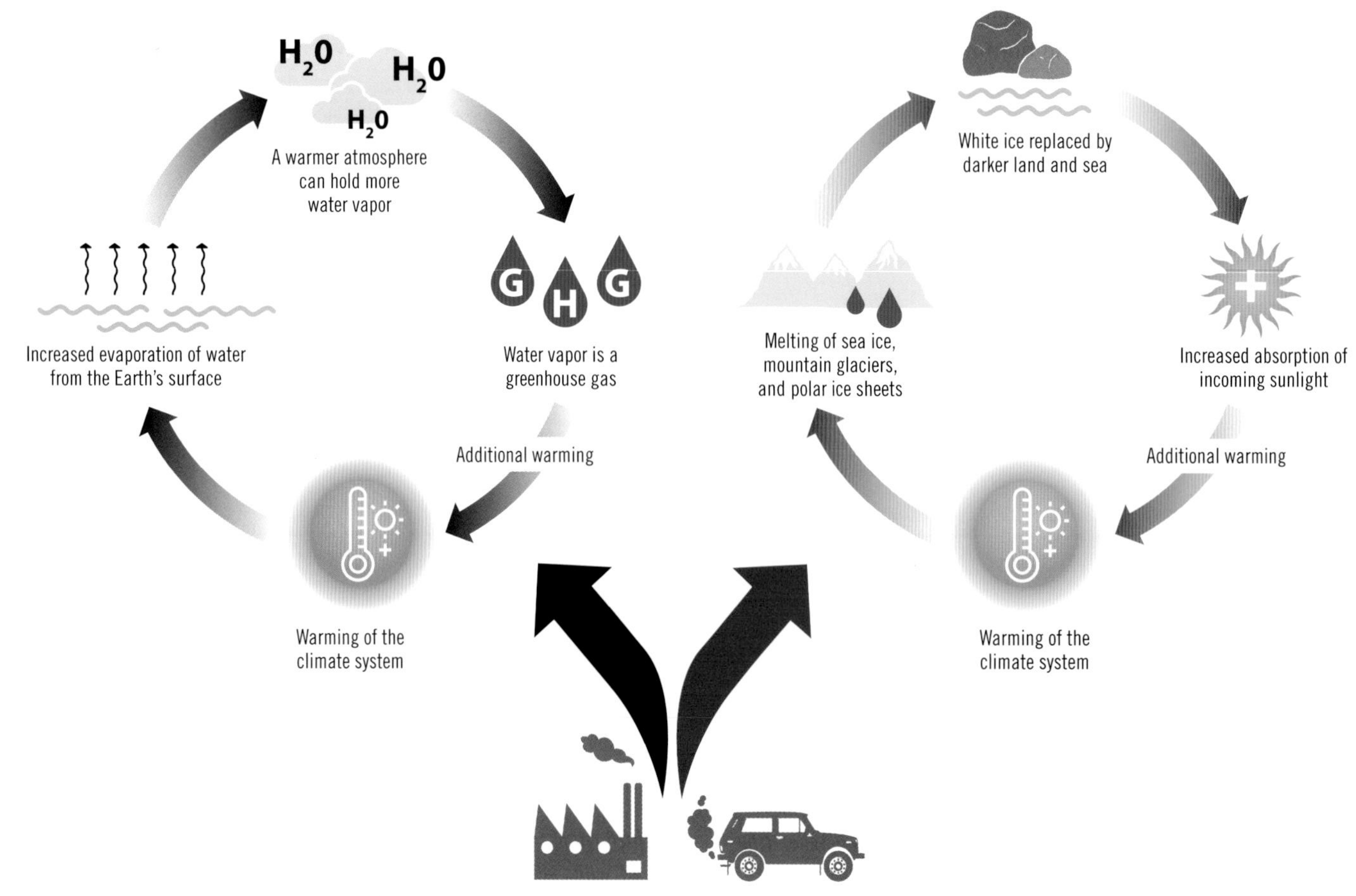

disruption of the ocean carbon cycle (see page 116) is a major concern. The oceans are the world's largest carbon sink, but as ocean waters warm, the efficiency of the sink is being reduced, increasing atmospheric carbon levels and hence speeding up global warming.

One of the few possible negative climate feedbacks is what is known as carbon dioxide fertilization, whereby higher carbon dioxide levels stimulate plant growth, which would help to remove carbon from the atmosphere. Unfortunately, however, any stimulus may well be offset by factors such as high temperatures and drought that act to inhibit growth.

Clouds could act as either a positive or a negative feedback. A warmer world means more evaporation, which means more water vapor in the atmosphere, which should mean more clouds (although higher temperatures also tend to evaporate clouds). More clouds mean more of the Sun's energy is reflected back into space, which could slow warming. However, water vapor is a powerful greenhouse gas, so an increase in water vapor could accelerate warming. For these reasons, the impact of clouds on climate change remains a leading source of uncertainty in predictions of future global warming.

There are also fears that the Earth is approaching (and may even have passed) one or more so-called tipping points, when a small rise in temperature creates sudden, unexpected, and permanent changes with global consequences. Examples include the collapse of the West Antarctic ice sheet, thawing of permafrost, dieback of the Amazon rainforest, and shutdown of the ocean conveyor belt (see below). Some scientists are warning of the possibility of a "cascade" in which the breaching of one tipping point causes the breach of another, leading to a rapid escalation of changes to the Earth's natural systems.

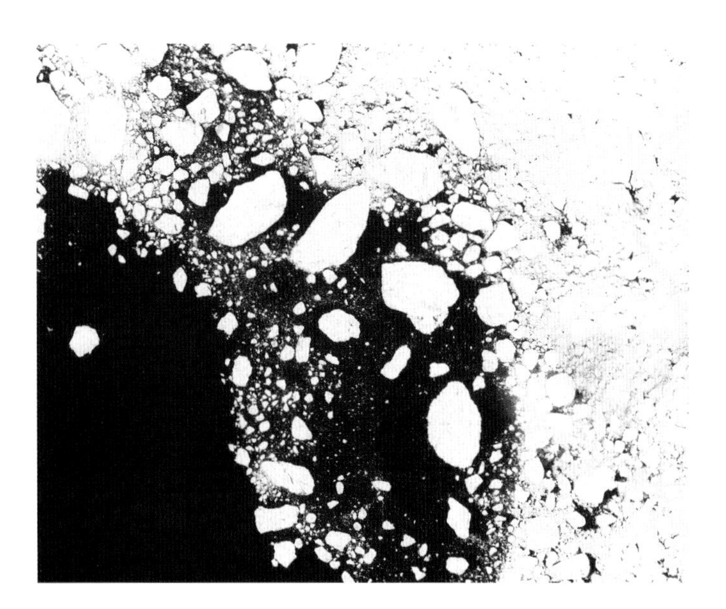

Above *Sea ice is highly reflective, while open water is very good at absorbing solar radiation. So, when sea ice melts, the exposed water warms, which causes more sea ice to melt.*

Below *Antarctic temperature trends, 1957–2006. Over the past 50 years, West Antarctica has been warming more rapidly than almost anywhere else on the planet, causing glaciers to retreat and ice shelves to collapse. There are now fears that the West Antarctic ice sheet is reaching a tipping point where its collapse will become irreversible, an event that would see sea levels rise by 11 feet (3.3 meters).*

Temperature change per decade (degrees Celsius)

0 0.05 0.10 0.15 0.20 0.25

THERMOHALINE CIRCULATION

Melting Arctic sea ice is reducing salinity in the Arctic Ocean and raising the water temperature, affecting the formation of the cold, dense water that drives the ocean conveyor belt. Recent research suggests that the conveyor has already weakened by around 15 percent since the middle of the twentieth century. If this continues, it could have a major impact on the oceans' ability to absorb carbon dioxide, while also significantly affecting the climate of the Northern Hemisphere, particularly Europe. It could even lead to the cessation of arable farming in the UK, for instance. It may also reduce rainfall over the Amazon Basin, cause near-permanent drought in Africa's Sahel region, disrupt the Asian monsoon and, by bringing warm waters into the Southern Ocean, further destabilize ice in Antarctica and accelerate global sea-level rise.

Impacts

Above *Polar bears use the Arctic sea ice for hunting and long-distance travel. As it disappears, they are forced to spend more time on land, which has a negative impact on their health and reproduction.*

The evidence that the climate is changing is all around us. One of the more predictable impacts of global warming is melting ice. Arctic sea ice is declining at more than 12 percent per decade and summer sea ice is projected to essentially disappear within 20–25 years at the current emissions rate. The Antarctic has been losing about 148 billion tons (134 billion tonnes) of ice per year since 2002. Mountain glaciers are also receding at an alarming rate, losing 6.7 trillion tons (6.1 trillion tonnes) of ice over the same time period. In total, since 2010 global annual ice loss has been roughly 1.3 trillion tons (1.2 trillion tonnes).

The melting of sea ice is destroying a key habitat for Arctic animals such as polar bears, seals, and walruses. Similarly, the melting of sea ice in the Antarctic has seen the Adélie penguin population on the western peninsula collapse by 90 percent or more.

Rising temperatures are also causing the Arctic permafrost to thaw, damaging houses, roads, and other infrastructure due to subsidence. In parts of Alaska, thawing permafrost has caused the ground to subside more than 15 feet (4.6 meters).

Most of the water from the melting ice eventually reaches the sea, causing sea levels to rise, a phenomenon exacerbated by the thermal expansion of warmer sea water. Sea levels have already risen by about 10 inches (25 centimeters) since 1880 and are continuing to rise by about 0.13 (3.3 millimeters) a year. Rising sea levels will increase coastal erosion and flooding, as well as saltwater intrusion into rivers and freshwater lenses on coral islands. Some low-lying islands will be completely submerged. If the Greenland ice sheet melts completely, sea levels would rise by about 23 feet (7 meters).

The addition of extra heat to the climate system is also wreaking havoc on the weather. Changing precipitation patterns are increasing the frequency and severity of drought; extreme rainfall events are becoming more common, too, with growing frequency and severity of floods.

Although the current consensus holds that a warmer world will not necessarily mean more tropical cyclones, we are already seeing a rise in the occurrence of the most powerful, and hence most destructive, storms. By altering large-scale atmospheric circulation patterns, global warming is also shifting storm tracks from their typical paths.

Because the polar regions are warming more quickly than the rest of the world, the temperature contrast that drives the jet streams has decreased. Slower, weaker jet streams have been linked to melting in Greenland and a potential rise in deadly weather events since they can lock weather systems in place, stalling them over regions. Weaker jet streams also increase the likelihood that Arctic air will escape the polar vortex (see page 176) and move south, bringing frigid weather to Europe and North America.

Changes to the weather have many flow-on effects. Longer dry spells, as well as more frequent periods of

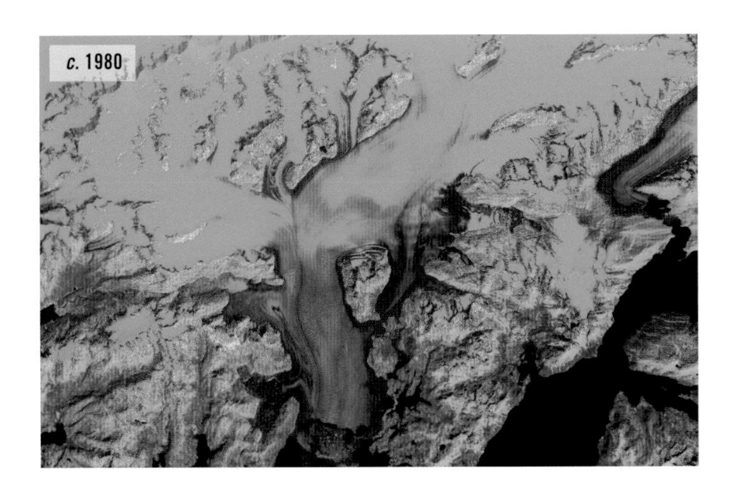

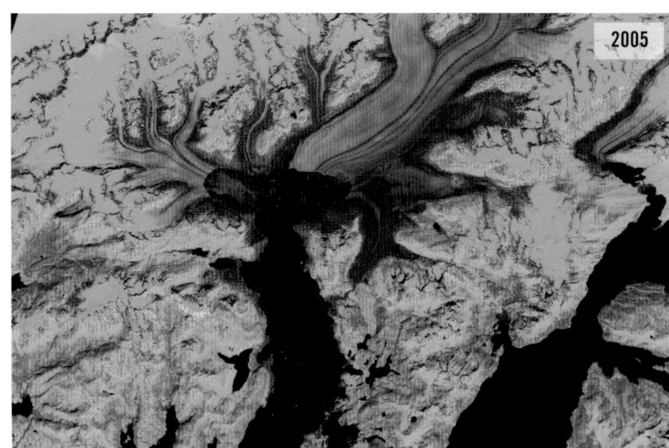

Above left and right: *The Columbia Glacier in Prince William Sound in Alaska, photographed in around 1980 and 2005. Since 1982, the glacier has been retreating at a rate of about 1,968 feet (600 meters) per year. It has also been thinning significantly, losing more than half of its total thickness and volume since the 1980s.*

extreme heat, are increasing the prevalence and severity of wildfires. Changes in the timing of rainfall and its growing unpredictability, an increasing frequency of drought and extreme rainfall events, and rising heat stress are all having a detrimental impact on agriculture in many areas, threatening regional and global food security, and pushing many farmers into poverty. The greater frequency of heatwaves is killing the old and infirm. Warmer temperatures are also lengthening the pollen season and generally worsening air quality, both of which can result in more allergy and asthma attacks, a decrease in lung function, and a worsening of chronic lung diseases.

Rising temperatures are having a direct impact as well, particularly on animal and plant species. Increasing water temperatures are devastating coral reefs around the world and threatening their future. In 2016 and 2017, extreme heating of waters surrounding the Great Barrier Reef killed roughly half its coral. According to a recent study, about 70–90 percent of all existing coral reefs are likely to disappear in the next 20 years.

In many regions, rising temperatures are allowing pest species to thrive. For example, booming populations of bark beetles, which feed on spruce and pine trees and are normally kept in check by cold winters, have devastated millions of acres of forest in North America. Disease-carrying mosquitoes are extending their range, bringing infections such as malaria, dengue fever, and West Nile virus to new regions. Pathogens such as the chytrid fungus, which has already caused amphibian extinctions all over the world, are also spreading more widely.

Animal and plant distributions are changing, both on land and in the oceans. Many terrestrial species are moving to higher altitudes and latitudes to find more suitable temperatures. However, many species cannot move, putting them at risk of extinction. In the oceans, changes to the ranges of both predators and prey are having knock-on effects: species such as puffins and penguins are having to travel further in search of food for their young as warming waters have caused prey populations to move; in Australia, sea-urchin populations have expanded, driving giant kelp forests to extinction.

The timing and patterns of animal migrations, particularly those of many bird species, are also changing as spring and fall arrive earlier and later respectively. This is having consequences as the timing of migration becomes disconnected from prey availability, so parents struggle to find food for their young.

Below *Flames engulf a road near Bastrop State Park, Texas, in September 2011. Rising temperatures are leading to an increase in both the frequency and intensity of wildfires.*

Further reading

The Cloudspotter's Guide by Gavin Pretor-Pinney, Hodder & Stoughton, London, 2006.

The Dictionary of Physical Geography by David S.G. Thomas, Wiley-Blackwell, Hoboken, 2015.

Eric Sloane's Weather Book by Eric Sloane, Dover Books, Chatham, 1952.

Essentials of Oceanography: International Edition by Alan Trujillo and Harold Thurman, Pearson, London, 2014.

Fundamentals of the Physical Environment by Peter Smithson, Ken Addison and Ken Atkinson, Routledge, London, 2008.

Fundamentals of Weather and Climate by Robin McIlveen, Oxford University Press, Oxford, 2010.

How the Ocean Works: An Introduction to Oceanography by Mark Denny, Princeton University Press, Princeton, 2008.

Introducing Oceanography by David Thomas and David Bowers, Dunedin Academic Press, Dunedin, 2012.

Introducing Physical Geography by Alan H. Strahler, Wiley, Hoboken, 2013.

McKnight's Physical Geography: A Landscape Appreciation by Darrel Hess and Dennis G. Tasa, Pearson, Cambridge, 2013.

Oceans: Exploring the Hidden Depths of the Underwater World by Paul Rose and Anne Laking, BBC Books, London, 2008.

Ocean: An Illustrated Atlas by Sylvia Earle and Linda Glover, National Geographic, Washington DC, 2008.

Physical Geography: The Global Environment by Joseph A. Mason, James Burt, Peter Muller and Harm de Blij, Oxford University Press, Oxford, 2015.

Understanding Weather and Climate by Edward Aguado and James E. Burt, Pearson, Cambridge, 2013.

Weather: A Visual Guide by Bruce Buckley, Edward Hopkins and Richard Whitaker, Firefly, Toronto, 2004.

Index

Picture credits

t = top, b = bottom, l = left, r = right

Alamy: 124t, 155tr

David Woodroffe: 54, 73t, 89b, 94, 109b, 151b, 162, 172

ESA: 103b, 142, 178

Flickr: 155br (Alan Stark)

GEBCO: 95

Getty Images: 37t, 39

GRID-Arendal: 19b (Peter Prokosch), 99r

IRI: 165b (Earth Institute/Columbia University)

NASA: 38, 44r, 68, 80t, 86, 89t, 108, 113, 114, 115b, 126, 130, 132, 137, 139b, 140r, 141tr, 143t, 151t, 157t, 164t, 169b, 174, 175b, 177b, 180b, 183b, 187tl, 187tr

National Archives and Records Administration, USA: 135

National Science Foundation: 80b

National Snow and Ice Data Center, University of Colorado, Boulder: 44l

NOAA: 81tl, 97, 98, 99 tl, 99bl, 101r, 176

Pexels: 74 (Boriz Ulzibat)

Science Photo Library: 10, 11, 16, 30, 75tl, 93, 100, 105, 111, 159t, 159b, 171

Shutterstock: 2, 4, 5 , 6, 7, 8, 13, 14, 15t, 17, 18, 19t, 20, 21, 22, 23t, 23b, 25b, 26, 27t, 27b, 31t, 31b, 33, 35, 36, 37b, 43, 45, 50, 51t, 51b, 55, 56, 57, 58, 59t, 59b, 60, 61, 63t, 63b, 64, 65t, 65b, 67tl, 67tr, 67b, 69bl, 69tr, 69br, 70, 71, 72, 75 bl, 75 tr, 77b, 78, 79t, 79b, 81bl, 81br, 82, 84, 85t, 85b, 87, 88, 91b, 103t, 106t, 109, 110, 117t, 118, 121t, 122, 122t, 123b, 124b, 125l, 125tr, 127, 128, 131, 134t, 136, 138, 146, 147, 149t, 154, 155l, 156b, 157b, 160, 163, 164b, 169tl, 169tr, 173, 179tl, 179tr, 179b, 180t, 182, 183tl, 185t, 186, 187b

Shutterstock Editorial: 120

Wikimedia Commons: 24, 25t, 28, 29b, 34, 41, 42, 46t, 46b, 47, 48, 49t, 49b, 52, 53, 62, 66, 76, 91t, 92, 101l, 102, 106b, 107, 112, 117b, 121b, 125br, 133, 136r, 139t, 145, 148, 153, 158, 167t, 167b, 175t, 185b

World Resources Institute: 161